ENERGY S

My Experiments
with SOLAR Truth

Chetan Singh Solanki

notionpress
.com

INDIA · SINGAPORE · MALAYSIA

Notion Press

Old No. 38, New No. 6
McNichols Road, Chetpet
Chennai - 600 031

First Published by Notion Press 2019
Copyright © Chetan Singh Solanki 2019
All Rights Reserved.

ISBN 978-1-64650-945-4

Dedicated to

MAHATMA GANDHI
His Ideals & Innumerable followers world-wide

Edited by
Bienu Verma Vaghela

Sometimes...

In our limited vision, we miss out the bigger picture,

In our ability to understand only finite, we forget the infinite,

In our limited wisdom, we tend to make mistakes much harmful than we think,

In our limited capacity to comprehend, we cannot understand the vastness of the universe,

In our one-dimensional focus on wealth creation, we forget the multi-dimensionality of life,

In our desire to make life comfortable, we forget the discomfort that we cause to others,

In our short life span, we forget the timelessness of the Earth.

– Chetan Singh Solanki

CONTENTS

Contents

PREFACE

Gandhiji has always been a strong influence on me, my thoughts and actions, even the course of my life takes a leaf out of Gandhiji's inspirational & incredible life. Having read quite a bit of his works, today I can safely say that I am in awe of his writings and teachings, particularly his autobiography – The Story of My Experiments with Truth.

Only after reading it a few times, I started to understand how Gandhiji's experiments provide us hints on designing a sustainable future for ensuring the long-term existence of human lives on Earth.

Many of my SOLAR experiments, especially during the latter part of my career are inspired by and are in line with the Gandhian philosophy, leading to what I call "Energy Swaraj." Therefore, I am daring to title this book as Energy Swaraj - My Experiments with SOLAR truth. Only time will tell whether justice has been done to this title.

"Well done humans. We have conspired to foul our own nest so egregiously that our extinction is now all but assured" wrote Steve Hanley in his article based on a scientific study published in 2018.

Our actions have ruined the world so much that, in business as usual scenario of energy generation and consumption, there is some finite possibility of human extinction from the planet, as early as by the end of this century. The use of fossil fuel-based energy is inflicting catastrophic climate change. The climate is changing day-by-day, faster than anyone of us would have anticipated. But most of the population in the world, unfortunately, does not even understand the severity of the issue and continue to live with business as usual way.

The current system of energy generation and transmission is a misfit in the definition of a sustainable planet. Using the fossil fuel-based energy is like digging our own graves, stealing future of our own children and exploiting our own home. We have become users of nature and forgotten that we are part of nature.

I am surprised even to realize that such energy ecosystems based on coal, oil and gas have even got established in the world. How can energy be transported thousands of kilometres, processed with the unwieldy and complex machinery and transmitted again to hundreds and thousands of kilometres for consumption, that too with very low efficiency? It only appears to be the short-sightedness of our forefathers and leaders of the world.

Humans have been on this earth for more than 200,000 years and I hope that all of us would love to live at least this much more. Fossil fuels, which have become an integral part of our lives, are in use only for a very short time say 100-200 years, and may last 100 - 200 years more.

In the above context, it sounds almost crazy to design an energy ecosystem which entirely runs, well almost, on fossil fuels. These are in limited quantities, unevenly distributed, and above all, responsible for climate change on earth. Have humans lost their mind to have designed, literally infinite existence of the human race on finite fuel?

Who would have thought that climate change would become such a big scare that people we will have to think about reversing energy solutions? None of us did that!

But yes, Gandhiji did, even in those times, over 100 years ago, who envisioned "Gram Swaraj" which propagated production and consumption of all goods locally, providing some major ground for sustainability. Likewise, "Energy Swaraj" would lead the way for localisation of energy generation and consumption for sustainability.

If the book, which is derived from my actual experiments, first-hand experiences, failures and successes with Solar, wakes you, shakes you, alarms you, startles you, interests you, captivates you, inspires you and motivates you to take that first step towards mitigating climate change by opting cleaner, greener, renewable source of energy – Solar Energy, I have succeeded in my objective of writing this book.

– Chetan Singh Solanki

FOREWORD

This book is a refreshing vision for the future of our fragile planet. A very large chunk of the problems humankind is facing today is due to the manner in which we have plundered the resources of this Earth and among them, fossil fuels are the most corrosive.

Energy Swaraj, I believe, draws its inspiration from Gandhiji's Gram Swaraj, which had relevance then, now and in the future for humankind's sustainability. In the same way, I believe this book will go a long way towards the struggle to maintain sustainability on planet Earth in the future.

Energy undoubtedly is a basic necessity. But when Gandhiji wrote, *"There is a sufficiency in the world to meet the needs of men but not the greed of any,"* he saw the felling of trees, the mining of coal, the damage to our rivers, and condemned rampant industrialization, but not being a scientist himself he did not write about alternatives.

Instead, he chose to move out of the city in 1904 in South Africa and lived close to nature at the Phoenix Settlement where there was no electricity, no piped water and no municipal services. Indeed, if he had solar technology at that time, he would have embraced it most ardently I believe.

In his book, Satyagraha in South Africa, he wrote about life on the Tolstoy Farm, his second home away from the City, also in South Africa, *"In spite of a large number of settlers, one could not find refuse or dirt anywhere on the Farm. All rubbish was buried in trenches sunk for the purpose. No water was permitted to be thrown on the roads. All wastewater was collected in buckets and used to water the trees. Leavings of food and vegetable refuse were utilized as manure. A square pit one foot and a half deep was sunk near the house to receive the night soil, which was fully covered with the excavated earth and which therefore did not give out any smell...*

...There were no flies, and no one would imagine that night soil had been buried there. If night soil was properly utilized, we would get manure worth lakhs of rupees and also secure immunity from a number of diseases. By our bad habits, we spoil our sacred river banks and furnish excellent breeding ground for flies with the result that the very flies which through criminal negligence settle upon uncovered night soil defile our bodies after we have bathed."

He did not refer specifically to energy but he was indeed concerned about the environmental effects of the use of fossil energy and other resources of the world when he spoke about us living as *"a part of nature and not apart from nature."* I would guess that the use of Solar energy would be living as a part of nature.

In its propagation of Solar energy, this book promotes self-reliance. Like Gandhiji's Gram Swaraj this book encourages the usage of scientific techniques of conservation to enable communities to become self-reliant. But the book goes on to plead for behavioural changes for the successful mitigation of climate change by encouraging people to adopt cleaner, greener energy, such as Solar energy.

I believe like Gandhiji's Gram Swaraj, Energy Swaraj would have world-wide appeal and application. Adopting Solar energy will shed people's dependence on Governments and big businesses and will result in people becoming independent. Adopting Solar energy and becoming conscious of issues such as climate change, rising pollution and other related issues, the world will be able to successfully reverse the rapid environmental degradation.

The book's reference to reading using Solar lamps is an encouragement to learning and resonates closely to Gandhiji's encouragement to the youth to read good literature and learn from it. He himself was an avid reader. His collection of reading material clearly indicates that the choice of books was an important part of learning. The market is saturated with books that promote hatred, violence, immorality and all the other social ills.

We need books such as this book which teaches us to conserve, to learn about alternate technology and to protect our environment.

So, this book and its advocacy for the use of Solar energy would I think be a most befitting testimony on the 150th birth anniversary of Mahatma and Kasturba Gandhi.

– Ela Gandhi

Author, Peace Activist and Granddaughter of Mahatma Gandhi

Member of Parliament, South Africa (1994-2004)

ACKNOWLEDGEMENT

The book in your hands: Energy Swaraj - My Experiments with SOLAR truth transcends you through the journey of 45 years of my life, particularly of 20 years of my experiments in the field of Solar. Many people have contributed immensely in my Solar journey; it is my great pleasure to acknowledge their contribution and support.

I am lucky to have very hard-working, kind and humble parents, Maa and Papaji, and my gratitude, love and respect for them are beyond words. My brothers Shiv, Kuldeep and Rajendra have also played a significant role.

The strength of my character and capacity to envision life beyond the visible horizon is provided by Gurudev Sri Sri Ravi Shankarji's Art of Living. My special gratitude for him.

I am grateful to Prof. J. Vasi, Prof. R. Lal and late Prof. A. Chandorkar for shaping up my technical skills. I am also thankful to Prof. J. Poortman, Prof. J.P. Celis, Dr. R. Bilyalov, Dr. G. Beaucarne from Belgium for guiding and helping me in my initial experiments in the Science and Technology of Solar cells, while I pursued my Doctorate in Belgium.

I am indebted to Prof. P. Vyavahare who influenced me a great deal not only during my engineering studies but till today. I thankfully acknowledge the contribution Prof. N. C. Narayanan, but specifically the contribution of Prof. Jayendran V. who has been more than a colleague and great fellow in many SoUL experiments. I am thankful to Dr. B. Ranga for some great discussions and help on Solar lamps.

My special acknowledgement and thanks for the administrative support provided by IIT Bombay, special thanks to former Director, IITB, Prof. Devang Khakhar and Deans Prof. P. V. Balaji, Prof. K. P. Kaliappan, HoD Prof. R. Banerjee. I am thankful for the guidance and support provided by Prof. D. B. Phatak on various occasions. Several discussions and administrative support from colleagues like Prof. B. Chakravarthy, Prof. Wagle, Prof. Raja, Prof. Surya, Prof. Pratibha, Prof. Anish, Prof. Karthik are greatly acknowledged.

I am thankful to Mr. Bharat Jhawar, with whom I started the Solar lamp experiments which laid the foundation for the work ahead. I also learned many things from Mr. Jiten Prajapati who designed the Solar-Passive architecture of Education Park.

In my admiration and respect, I am thankful to Dr. Anil Kakodkar for his encouragement and guidance and to Prof. Satish Agnihotri, who has been guiding me in difficult times. My special thanks to Dr. U. Tripathi, DG, International Solar Alliance.

My gratitude to Mr. Piyush Goyal, Minister of Railways & Commerce for his fair and quick assessment of our Solar lamp program and helping it to grow beyond, which enabled me to perform many of my Solar experiments. The support from former Secretary, MNRE Mr. G. Pradhan, Mr. D. Gupta and present Secretary Mr. A. Kumar has been encouraging. I also thank officials from the Government of India including Mr. Santosh V., Mr. Tarun Kapoor, Mr. G. Gupta, Dr. P.C. Maithani, Dr. G. Prasad, Mr. J. Jethani, Dr. B. Bhargav and Mr. S. Garnaik.

The contribution of many SoULS team members, particularly from Ms. Abhilasha, Keyur, Purva, Fanny, Felix, Jivita, Pavan, Sagar and many others is immeasurable.

Acknowledgement

I am thankful to every member, especially to the Gandhi Global Solar Yatra team, dynamically led by Ms. Swati Kalwar, who has worked amazingly hard to make this yatra a grand success with the invaluable support of Nikita and Harshad, supported by team members Vinit, Ameya, Meghen, Pankaja, Kashyap and Jagdeep, working untiringly for several months.

Mr. Auroshis Rout and Dr. Vishnu Kant Bajpai have a special mention here for devoting their time on various discussions, writing assistance and data inputs for the book.

My special thanks for the Ministry of New & Renewable Energy (MNRE), Government of India, Tata Trust and Idea Cellular for funding SoULs project. My special thanks to Ms Amita Sharma and Dr. Nilay Ranjan from Idea Cellular for their timely support.

However, the most important contribution is by my wife, Rajni, who has been a great support system and has worked extremely hard to see where I am today. My lovely daughters Suhani and Mahak are great "energy recharge" for me always.

Lastly, I am thankful to Ms Bienu Verma Vaghela, Editor, for meticulously anchoring and carefully editing the book and constantly chasing me to write until I completed writing. I am also thankful to Sarbani for proofreading.

There are many others who have contributed to my work. I acknowledge the contribution of one and all who have been part of my journey.

– **Chetan Singh Solanki**

Illustrations Courtesy: Prof. Arun B. Inamdar

*When I admire the wonders of a sunset or the beauty of the moon,
my soul expands in the worship of the creator.*

— Mahatma Gandhi

PART - A

ENERGY IS EVERYTHING,
EVERYTHING IS ENERGY!

Dear Sun,

> *Without you, the trees will not grow,*
> *Without you, the wind will not blow,*
> *Without you, the river will not flow,*
> *Without you, the seasons will not billow,*
> *You shine bright and show us the way,*
> *Oh SUN! You are our nourishing STAR, always!*

CHAPTER 1

THE WORLD IS A "PLAY OF ENERGY"

यदा यदा हि धर्मस्य ग्लानिर्भवति भारत।

अभ्युत्थानमधर्मस्य तदाऽऽत्मानं सृजाम्यहम्।।4.7।।

परित्राणाय साधूनां विनाशाय च दुष्कृताम्।

धर्मसंस्थापनार्थाय संभवामि युगे युगे।।4.8।।

Yada Yada Hi Dharmasya, Glanirva Bhavathi Bharatha,

Abhyuthanam Adharmaysya, Tadatmanam Srijami Aham// 4.7//

Praritranaya Sadhunam, Vinashaya Cha Dushkritam,

Dharamasansthapnaya, Sambhavami Yuge-Yuge // 4.8//

—Bhagavad Gita

This verse from ancient scripture Gita means that "Whenever righteousness (*Dharma*) declines in the world, whenever evil (*Adharma*) arises, I take birth in your soul. For the protection of the good and for the total destruction of evil, and to establish or re-establish *Dharma*, I take birth again and again."

Inspired by this verse from the Gita, I interpret that the prevalent energy scenario in the world today as: Whenever the dark energies (read *Adharma*) takes over the world, whenever destructive energy takes over the good energy and disturbs the balance of life on the Earth, nature somehow finds ways and means to restore the balance or restores the good energies (read *Dharma*). The time has come to sound mankind and inspire taking action towards arresting these bad energies overtaking

3

good energies in order to maintain the balance of life on the Earth.

Energy is a very important aspect of our lives. Energy is an important reason for the existence of the Universe, for the Earth, for every living being on the Earth, and that of every human being as well.

The world is a play of energy: An amalgamation of positive and negative energy, creative and destructive energy, righteous and unjustifiable energy, dark and bright energy...

...Coal energy and Solar energy!

In this modern world, energy is the main driver of social and economic growth. Today, every moment of our life and every square inch of space that we utilise has a touch of energy. It is really unthinkable to live a single day without consuming any form of energy. For individuals, higher use of energy not only enables better comfort but also higher income and better health. For an industry, higher energy consumption means more production of goods and services. And, for a country, higher energy consumption means higher Human Development Index (HDI), a larger production and higher GDP.

Humans have existed on this planet since several hundreds of thousands of years. It is only obvious that the energy source that humans consume should have also existed since long and should have the potential to exist for millions of years, for hopefully so much longer human existence on this planet.

However, today, most of the human activities, around the world, is running on finite energy resources – fossil fuels i.e. coal, oil and gas which are incessantly emitting Carbon Dioxide. Also, human activities based on fossil fuel consumption are causing enormous pollution on planet earth. Not only this, it is creating a huge imbalance in nature, which is leading to climate change and making sustainability of human beings on this planet questionable.

Surprisingly, the Sun which has been shining for millions of years and is the main driver of the existence of life on the planet Earth has not been considered to it's full potential yet, though it is the most feasible option to counter the scenario. The Sun has been nurturing and nourishing the lives of millions of species from millions of years, not only this, it's energy provides the required warmth on the planet and enables plants to grow.

Its energy gets manifested in different forms and enables rivers to flow and air to blow. For thousands of years, humans have learned to harvest these energies for running their lives, for obtaining required comfort, growing and maintaining food, for transportation, for cooking and for sailing.

Till as late as 1850, we see that the life on the planet with regard to energy generation and consumption was in complete sync with nature. Gradually, the scenario began changing, the dark energies of coal and crude oil made their way and soon gained momentum. With the onset of the industrial revolution (since the 1850s), we discovered the use of coal and oil. The balance of energy consumption and generation on the planet started getting disturbed.

Since then, we have been digging out the Carbon stored beneath the Earth's surface and started burning these fuels. Such activities have started emitting Carbon in the form of Carbon Dioxide in the atmosphere. Prior to 1850, the exchange of Carbon Dioxide between the Earth and atmosphere, i.e. emission to the atmosphere and absorption by the atmosphere, was in the right balance, therefore the existence of life was in balance.

Since industrialization, humans have changed. Instead of being part of nature, we have become users of nature. We do not feel being part of it. How otherwise, we would keep emitting smoke in the environment, keep cutting the trees, keep polluting the rivers, keep degrading the soil and keep throwing the garbage? The human greed to acquire better comfort and accumulate greater wealth has resulted in air pollution, water pollution, soil erosion, deforestation, sea-level rise and melting of polar ice.

Moreover, since the onset of the use of fossil fuels, the emission of Carbon Dioxide is much more than its absorption from the atmosphere. This has tilted the balance of nature in such a manner that it increased the amount of Carbon Dioxide in the atmosphere, causing the increased greenhouse effect. This causes severe and catastrophic climate change, so much so that now it is threatening the very existence of human life on the planet.

Isn't this situation scary?

Hence, I felt that there is a need to spread energy literacy to bring awareness on the negative aspects of dark energies (coal, crude oil) and positive effects of the good energies (renewable sources). There is a need to add a sustainability dimension to energy generation and consumption for the long-term existence of humans on the planet. There is a need to add a moral and spiritual dimension to energy, on top of the only economic dimension of energy that exists today.

CHAPTER 2

ARE WE SINKING INTO A DEEP, DARK WORLD OF ENERGIES?

Energy, this is the term I am sure everybody has heard in his/her life, in one way or the other. Most of you may have used this term many times casually referring to your own body or any machine or gadget you are using, without even knowing it clearly, what it is or what it implicates.

You may not clearly understand its mechanical implications but physical implications yes, when we say, *"Today I am feeling energetic or today I am feeling a total loss of energy."*

This is energy which keeps us going, enables us to carry out all the day-to-day activities, if there is a loss of energy in the body, we feel weak and are unable to carry out any activity. If there is total loss of energy, we may collapse.

So, what is energy?

Energy is a prerequisite to run not only the human body but also all machines on the planet, wherever there is an activity, there is energy. It is omnipresent 24 x 7, running all living beings and machines on the planet.

You would have come across the terms like: Energy supply, Energy shortage, Energy security, Energy purchase, Energy bill, Solar energy, Renewable energy, Energy conservation, Energy access, Clean Green energy, Coal energy, Nuclear energy, Hydro energy, Light energy, Heat energy, Chemical energy, Mechanical energy, Electrical energy and many others. This way we all are surrounded by the energy in a literal sense.

Does this affect our lives? Should we as a common man really think about energy, or later as it would turn out, shall we really worry about energy?

Also, you might have heard about the law of Conservation of Energy, which states that energy can neither be created nor destroyed, it can only change form from one to another.

Does this statement make things related to energy more complicated? Probably it does!

Like God, energy has no beginning, no end. Both are all-pervasive, there is no place on the planet where energy is not there, in some form or the other, it is present just like God. We have known that Energy and God have been around since creation and will be there till eternity.

Both God and Energy, make the entire world go round. Nothing moves without the power of energy. Everything around us is the play of energy. Positive energy makes us feel good, and negative energy makes us feel bad.

Coming back!

Fortunately, or unfortunately, we all are using energy in our daily lives. But can we describe it? Have you ever seen energy? Have you ever smelt energy? Have you ever touched energy? No, we can't see it, touch it or feel it, we can only see the manifestation of the energy.

Energy is all around us. The shining of the Sun is energy, the blowing of wind is energy, the playing of music is energy, the falling of rain is energy, blinking an eyelid is energy, hence every action that happens within us or around us is energy or as they say, the play of energy. If a person or an object performs some work, the object or person is said to possess energy.

Every action that happens around us manifests the use of energy. The use of energy lights our homes, powers our vehicles, runs our fans, cooks our food, cultivates our fields, runs our factories, runs our television, runs our motors, and everything else where some other action is involved, this manifests use of energy.

Are you getting a sense of energy?

NO MINUTE PASSES WITHOUT A TOUCH OF ENERGY

The energy which we are using can either be derived by using our own body or it can be derived from external sources. In modern times, we are using less energy from our body and more energy from external sources. Since industrialization, especially over the last century and more, our life has become very mechanized. We are using machines to do more work for us than using our own body, which means the more use of energy.

Since the onset of industrialization humans have learnt to use and derive energy from various other non-living sources like coal or oil, gas or hydro. Slowly, we started using machines in the factory to do work, we started using engines to drive us to distances and we started using electricity for lighting and for various other purposes.

We are increasingly becoming dependent on the energy from external sources, so much so that every minute of our life whether we are awake or sleeping, we consume energy in one way or the other.

The machines or gadgets have become such an integral part of our life, that we do not even realize that *whether we are using energy or misusing energy!* It is not an exaggeration, but I believe, this lack of awareness is causing climate change.

Machines are playing a highly significant role in our daily lives, so much so that life looks impossible without all these energy consuming machines. Men invented machines to make life simpler, but as time moved on and technology advanced, the devices which were created for practical reasons graduated to fancier levels.

Now you think of anything on this planet, it is associated with a gadget or a machine. Earlier we controlled the gadget, now gadgets control us. Earlier people used to start and end their day looking at the faces of their loved ones, now for the majority of them, the day starts and ends with a gadget.

Let me give you an example of the use of energy required in a simple activity like brushing your teeth every morning. Before I start, hold on for a minute to guess how much energy would be used for brushing your teeth. I bet you can't guess, so let me illustrate here.

This small activity of brushing your teeth requires a lot of logistical arrangements before it is ready for use. Like somebody has manufactured toothpaste in a manufacturing unit, packaged it, transported it to various outlets, stored in go-downs for days or months before we buy it. So, before we use it, just see, only toothpaste has to pass through so many stages and all this consumes energy.

A similar procedure is required for a toothbrush, mirror, tap, washbasin, etc. Manufacturing washbasin and mirror is a very tedious and energy-intensive process. From manufacturing, transporting, storing, cutting and fitting it in a frame and fixing it in your bathroom is a lot of work, which involves a significant amount of energy.

Brushing teeth also requires water which many times comes from hundreds of kilometres or drawn from hundreds of feet down the ground which requires a lot of energy to pump in the overhead tank, where it gets stored before you use it for brushing your teeth.

Now you can imagine what would be happening during the entire day when you are eating, bathing, drinking, cooking, cleaning, watering plants, driving to the office, working and sleeping in the night under the fan or AC? Each operation is using energy in several ways. So, whether we are awake or asleep, we end up using energy, every single moment!

NO SPACE AROUND US IS WITHOUT A TOUCH OF ENERGY

Not only that every minute we use energy but every space around us has a touch of energy. In the space around you alone there are many objects, where each object must have consumed some energy before it is put to use, in the shape and the place where it is operational now.

In technical terms, this is embedded energy. It means the energy that is in the object by the way of manufacturing or transporting for being in the place.

Look around, where you are sitting right now. If you are sitting in a room, you will see furniture, gadgets, floor and all other things as part of the space. If you are outside the house you would see buildings, roads, vehicles, divider, signboards, gardens, etc. Each of these have come from somewhere, each of these have used some amount of energy before they have come to the shape and size you are seeing them in right now.

If it is a wall, energy has gone into the manufacturing of all the materials, their transportation to the site, before it was erected like a wall in the building. If it is the paint on the wall, it had been prepared and transported before it was used for painting the wall, if it is furniture, somebody has made it for the use in required shape and size. If you see a road, you will notice that a lot of energy is used in making the foundation of the road, putting tar on it and rolling it over, then only it becomes useful for commuting. This way, we see that energy is used for all human and mechanical activities.

In my house, I could not find a single square inch of space, which has not utilized energy in one form or the other. I am sure this would be true for most of the readers as well. Else, take a pause, look around and observe the use of energy.

Not long ago, a few decades ago, in my village in M.P. India, there were many houses which were built and maintained by using the locally available soil and wood. I can go back in time and tell you that every square inch of the house was built and maintained without the use of any fossil fuel-based energy. Take note and become aware of this drastic change in the scenario, the sooner you do, the better it is.

Since, we do not see things around us in the form of energy, we are not sensitive towards the use of energy, we do not feel it, therefore we end up using a lot of energy. We even end up wasting a lot of it. Each unit of energy used, affects the environment and cumulative effect by all human beings on the environment has been so high in magnitude that it changed the very environment on the Earth which supports our lives.

Unbelievable? Isn't it? Does not matter. At least now, let's wake up to this fact.

Where does the energy come from, which we use? What are its sources? Well, our energy needs are fulfilled by using coal, oil, gas, nuclear and renewable energy sources like hydro, biomass and Solar energy. Today, in 2019, the unfortunate and worrisome part is that more than 80% of the energy needs of the world are fulfilled by the use of fossil fuels i.e. coal, oil and gas as shown in figure below. Since last few decades the trend remains nearly the same. We also know that fossil fuels are finite and their use causes a negative impact on the environment.

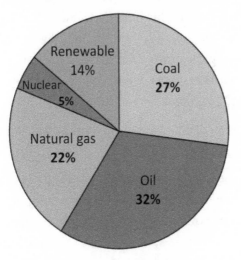

Figure: The energy sources in the world by fuel type

ENERGY CONSUMPTION FOR DIFFERENT END-USES

Now, we need to know how directly or indirectly we are consuming this energy. Direct use of energy is easy to understand, i.e. the energy that we use for cooking, running, travelling, etc. But unknowingly, indirectly, we are also responsible for a lot of other activities where energy is used. You may be the user of a big truck that carries oil, you may be responsible when a train carries coal, or you may even be responsible for a new shopping mall coming up in your area.

Think about it!

Technically, we use energy for domestic, industrial, transport, agricultural and commercial applications. Let us just go over all these applications.

In domestic applications, we use energy for cooking food, running lights and fans, cooling and heating of spaces, washing clothes, refrigerator, heating water for bathing, a motor for pumping water up in the tank, microwave and grill for food and ironing of clothes. We keep on adding many new devices for domestic applications like an air purifier, water purifier, dishwasher, clothes dryer, vegetable cutter, vegetable washer and what not! The list is far too long.

In our workspace we use a computer, laptop, charging mobiles, power banks, fans, ACs, Coolers, Printers, etc.

In industries, we use energy for running big motors, heating and melting materials, processing waste, water pumping, cooling, computing, air conditioning, lighting, ventilation and many applications. Many times, in industries the machines are of very high power which consume a lot of energy. It is not difficult to imagine very high-power consumption in industries, a lot of material which is very hard and long-lasting, is manufactured in industries.

How strong and rugged are railway tracks, ceramic washbasins, bridges built using steel, stainless steel vessels, etc.? One can imagine how much energy would be required to melt steel and other metals like aluminium, copper, etc. Imagine how much energy is required to melt glass to give the desired shape?

In transportation, be it goods or self, we consume energy for road travel, water travel and air travel. Besides, we need the energy to transport coal, petrol, gas and diesel. We need the energy to transport food items, goods produced in the factory and to transport billions of people every single day, many times in the comfort and time of one's choice. Particularly in road travel, we use two-wheelers, four-wheelers, trains, taxis, buses, metros, etc.

The amount of energy consumption in road transportation has been growing as the number of vehicles, size and distance covered are increasing. Traffic jams are choking cities, everywhere around the world, this way we see that a good measure of energy consumption is happening in transportation. More traffic jams indicate more energy consumption, Simple! Recently as a part of my Gandhi Global Solar Yatra, I have travelled to more than 30 countries, across continents; I have not seen a single city where traffic jams are not there.

Among various energy application sectors, the energy usage is almost equal in transportation, industrial, residential and commercial sectors (including energy usage in agriculture) as shown in the figure below. There is tremendous scope in energy saving in each of these sectors.

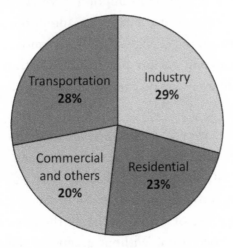

Figure: The use of energy in various sectors

The major transition in energy usage towards fossil fuel has happened mainly in the last few decades in most parts of the world. In my own life, I have experienced tremendous transformation in energy consumption. I remember, in the 1980s, when I was a child and studying in school, kerosene was the only external energy source that we used to buy for lighting purposes. Even to buy kerosene there was a struggle, we had to go to another village to fetch it. There was no LPG cookstove, there was no vehicle in the village, there was no fan, no AC and no TV.

Today, I live in IIT Bombay campus, which is as advanced and modern as campuses anywhere else in the world. The entire world around me has changed. All the electricity consuming gadgets surround me. In the campus, energy is consumed for doing anything and everything. Unfortunately, now I also had to struggle for energy, but for a very different reason; I had to struggle to choose a meeting room without an AC.

From Zero energy consumption several decades ago to living every minute with energy consumption, at home and outside the home, the change is really enormous. Isn't it? This enormous change in a short time coupled with the wastage of energy is leading to climate change, as many times it is just not need-based.

As a reader, are you getting a sense that how much energy usage has increased in the last few decades, and how you have been a part of it, directly and indirectly? Are you realising that we are sinking into a deep, dark world of energies? This drastic change is the cause of worry. Realizing the mistake is the first step towards correction. Without accepting this fact, we will never arrive at a possible solution.

By now you must be going crazy, how many different ways you consume energy or you are responsible for the consumption of energy. If not, I probably have not done justice with this chapter. We are all neck-deep in energy consumption, and unfortunately, more than 80% of the energy that we consume come from dark energies, the coal and crude oil*. Let's get out of this before it is too late.

Can we stop the use of all these deep, dark energies?

Ref : *www.ourworldindata.org

CHAPTER 3

NOT MONEY, ENERGY MAKES THE WORLD GO ROUND!

In modern times, all indicators of social and economic development are governed by energy. Several social and economic indicators of development have a very strong correlation with the quantum of energy consumption. Poverty status of a nation, it's literacy rate, it's per capita income levels, it's status of the economy, it's infrastructure, it's industrialization status and its people's quality of life, everything is governed by energy.

Indeed, energy is one of the most important parameters of development of any economy in the world. Energy has always played an important role in the lives of humans and the nations, as it acts as a parameter for both socio-economic development and an important humanitarian need.

Therefore, per capita energy consumption of a country is regarded as an important indicator of economic development. It is witnessed that social conditions do not improve until the per capita consumption of energy increases. Thus, developed nations show higher per capita energy consumption than developing countries.

This way, we see that everything revolves around energy, which is not only viewed as a production input but an important commodity, a commodity that affects international relations, an indicator of the prosperity of nations and even happiness quotient of people.

Energy is everything in a nation's life and the life of every human being. From the discussion, it is clear that the more energy we consume, the better off we are.

But the question is how much energy do we consume and how much energy should we actually consume?

But there is no clear answer for it yet. No scientist, no politician, no philosopher can say how much energy we should consume. You can come closer to the answer only when one defines the context. You should not be surprised to find very different answers if you look at it from the context of sustenance vis-à-vis the context of economics.

Let's first learn a little bit of maths, that we all need to know about energy.

MEASURE OF ENERGY

Since all the parameters of growth are measured in terms of energy, it is important to understand how do we quantify the use of energy. As a reader of this book and probably as a sensible energy user, the unit of energy and the magnitude of energy usage is important to understand. It will help you to put energy consumption on a per task basis, per capita basis and per country basis. The individual would be able to understand the lack of energy supply and overuse of energy consumption.

Let's understand the basic maths of energy as it is worth doing. There are many units of energy, I am putting here just one for the information. I would like to give you a feel of the quantum of energy that we use in terms of a commonly dealt unit of energy in our day-to-day life.

That is one kilo-Watt-hour or normally written as kilo-Watt-hour or kWh. You may not know the unit in this form (kWh) but all of us use this unit of energy. It is a basic unit of electricity used worldwide, and we pay the electricity bill in terms of a number of electrical units that we consume in a month. One unit is equivalent to 1 kWh of electricity. Though there are many units of energy, we will discuss all our energy consumption in terms of equivalent of kWh or electrical units.

Let's make some sense, how much is a unit of electricity or its equivalent energy. A 100-Watt bulb running for 10-hours consumes one unit of electricity, an instant electric geyser running for about 20 minutes would use one unit of electricity, a medium-size AC running for about 30 minutes would use one unit of electricity. A medium-size car can run about one km in one-unit of electricity or using petrol that is equivalent to one unit of electricity.

Have a look at your electricity bills for the last couple of months to find out the following:

- How many units of electricity you (your family) consume in a month?
- What is the cost of one unit of electricity that you are paying to the government?
- Is your electricity bill constant or it has been fluctuating?

Do not be surprised to note that several taxes are added in your electricity bill. Depending on where you are in the world, and for what purpose you are using electricity, your electricity rate varies from 5 to 40 USD Cents per unit or approximately INR 3 to 25 per unit.

The readers should note that sometimes the discussion is about the use of electricity and sometimes it is about the use of energy. When we say electricity, it only means the use of energy in the form of electricity. And when we say energy, it refers to the use of energy for all purposes like cooking, transportation, washing, industry, etc. in all possible forms like coal, petrol, diesel, LPG, etc.

ENERGY AND POWER

Also, it would be interesting for the reader to differentiate between the term 'energy' and 'power'. I have come across many people, including the educated ones, making a mistake in referring these two terms, energy and power interchangeably, which is not correct.

Is the term 'power' the same as 'energy'? No, they are not.

As a reader, do you really need to understand the difference? Probably not, but it is always good to know the difference.

Let me give you a simple analogy to understand the difference. The term 'energy' is referred to as something 'stored' or 'accumulated' or 'expended' over a period and 'power' is referred to something when the use of energy is in action.

In terms of stored water in the overhead tank and its use, the analogy goes like this. The 'energy' is stored water in a tank and 'power' is flowing water in a tap. Let's take the example of a hydropower plant: A dam stores water (stored water is energy) and when water runs through the turbine it generates power (flowing water is power), when petrol is filled in the car (stored fuel is energy) and when the petrol runs through the engine and gets burnt, it propels car (running fuel through a car engine is power), when the charge is stored in the battery (it is energy stored), when the same charge runs through a bulb, it lights it up which represents the power (flowing charge is power).

The bill for the amount of electricity used in a month (accumulated over a month, expended over a month) is paid in terms of energy, while the instantaneous consumption of electrical appliance is referred in terms of power, 40-Watt bulb, 60-Watt fan and 3000-Watt geyser, etc.

ENERGY AND STATUS OF A COUNTRY AND ITS PEOPLE

Generally, nations are classified as: under-developed, developing and developed, depending on the various developmental indicators among which energy is one of the most prominent one. Higher energy consumption in a country is an indicator of better economic and social status. Let's look at some of these indicators and their correlation with energy consumption.

The total energy consumed by a country is the sum of all types of energy consumption of individuals and industry including energy consumed in transportation (petrol, diesel and CNG), energy used in cooking (LPG) and energy used in household and workspace in the form of electricity. The total annual value of all economic activities that take place in a country is represented by Gross Domestic Product or GDP.

Is there any relationship between GDP and the amount of energy consumed in the country? Yes, very much, there is a direct relationship between GDP and energy consumption. Countries with higher GDP tend to consume more energy.

It is clear from the previous section that richer countries consume more energy, but does it also translate to higher per-person energy consumption in the developed countries? Do people in developed countries consume more energy annually and the people in the under-developed countries consume less energy annually? Yes, true.

By an estimate, for rich countries, the total energy consumption per person is more than 30000 electrical unit equivalent, for the moderately rich countries the per-person energy consumption is in range of about 15000 electrical unit equivalent while for the poor countries this number is only in a range of about 5000 electrical unit equivalent per person. The numbers are indicative and significantly vary within each category of countries. Take note of the huge difference in developed countries and underdeveloped countries.

Electricity consumption per capita gives a better indication of the well-being of people in a country. Having access to electricity at the household level improves the living conditions and quality of life significantly. There is a huge gap in the per capita annual electricity consumption within countries and between countries. For example, the per capita annual electricity consumption in developed countries is above 10,000 units, while the same in under-developed countries is just about 500 units or even less.

Generation of electricity and its supply requires more elaborate infrastructure as compared to the one required for coal, petrol, diesel and LPG. Electricity needs to be produced instantaneously by power plants to fulfil a real-time need, it is generated and used instantaneously, it is generally not stored, whereas oil, coal and LPG can easily be stored.

So far, you have seen how rich countries tend to consume more energy overall, and also more energy on a per capita basis and how poor countries tend to consume less energy overall, thus consume less energy on a per capita basis.

The next question: Is there any relationship of energy consumption with social and economic indicators like income, education, health, etc.? The answer is 'Yes'. Energy consumption does affect these indicators as well.

Let's look at these relationships.

All the rich countries of the world with high electricity consumption have good per capita annual income in the range of 50,000 USD. On the other hand, poor countries with low per capita energy consumption has an annual income in the range of only 1500 USD. These numbers are indicative and presented here to give a sense to readers, whereas the actual numbers in each category vary significantly from country to country.

Globally, all the rich countries have a very high literacy rate, mostly above 99%. All poor countries with low electricity consumption have a lower literacy rate within a range of 72 % to 86 %. The middle countries hover in between.

Even women empowerment is related to the use of energy. In 2017, about 87% of rural households and even 26% of urban households were cooking on biomass*. Women have to spend significant time collecting biomass and use of biomass for cooking, which has some severe impact on their health like respiratory, pulmonary and vision problems.

We need to add energy to the list of basic physiological needs of human beings like food, water, clothing, shelter, etc. It is important to first fulfill these needs, before any other higher-level needs of life.

ENERGY IS THE GOLDEN THREAD

The UN Secretary-General Ban Ki-Moon once said that *"Energy is the golden thread that connects economic growth, social equity, and environmental sustainability."*

Generation and consumption of energy and its easy access to every individual is very important, however, not at the cost of environmental sustainability.

In the year 2015, 193 members of the United Nations Development Program (UNDP) formulated 17 Sustainable Development Goals or commonly referred to as SDGs to be achieved by 2030. Each goal focuses on a different aspect of human life. The vision was to end poverty, protect the planet and to ensure that everybody has peace and prosperity.

Amongst the 17 SDGs, goal number 7 or SDG-7 is very specific on energy supply. It emphasizes on four dimensions of energy access: affordability (cost), sustainability (renewable energy), reliability (good quality), and modern source (electricity).

Though all the seventeen goals focus on different agendas, the common thread between them is their alignment towards improving the quality of life, all of which are connected to access to energy.

The access to affordable, reliable and sustainable energy supply is key to all developmental goals, as energy impacts health, education, literacy, income, industrialization, gender equity and almost everything else in our social and economic realm. In 2017, there were still about 1 billion people in the world, who did not have access to electricity and a massive 2.8 billion people in the world did not have access to clean cooking fuels. The 2.8 billion people are nearly 40% of the world's population.** This data shows that we need to generate and supply more and more energy for all these people, who lack access to energy for their betterment and empowerment.

The access to energy and its use is important but more than 80% of the energy consumed in the world today is fossil fuel-based, which negatively impacts the climate. As per the sustainable development goal 13 on mitigation of climate change, we need to severely cut down on fossil fuel usage.

In this way, the sustainable development goal on climate change is a complete contradiction with all other sustainable development goals. Improvement in energy consumption helps to achieve other sustainable goals but with heavy fossil fuel usage, we will fail in achieving the goal of climate change. On the other hand, if we stop using fossil fuel-based energy, we may be successful in achieving climate change goal but we will fail in achieving other goals.

What a contradictory situation it is! We need to supply more energy on one hand and cut down the energy usage on the other.

I feel that it is one of the biggest contradictions human beings are facing today. Solving this contradiction and arriving at a mutually beneficial solution is a big puzzle and tremendous challenge. It is always desirable to produce and consume more energy, but not at the cost of the environment or sustainability.

If you are able to find a solution to this challenge, you would be worthy of nothing less than a Nobel Prize. Not just one, but many, as the scale of the problem is very big. So, the Nobel prizes are up for the grab. Solve the problem and get one!

I have personally experienced the gaps that exist between rich and poor, within a country and between rich countries and poor countries. I come from a rural area and I frequently travel to rural areas. Also, as a part of my Gandhi Global Solar Yatra, I have travelled to many rich countries like the US, Europe, UAE, etc. and to very poor countries like Uganda, Benin, Ethiopia, etc. Therefore, I have first-hand experience of the gaps which exist, particularly with regard to energy consumption across the world.

Looking at the energy gaps that exist in the world, finding a solution towards sustainable energy supply for everyone in the world, appears to be a near-impossible task, especially if we continue to rely on fossil fuel-based energy solutions.

I feel that some drastically different energy solutions are required. What would be these energy solutions? What shape and size these should it take? These are the biggest questions of the millennium!

Ref : *www.indiaenergy.gov.in
       **www.iea.org

PART - B

THE BEGINNING – TRAINING ON SOLAR ENERGY

Strength does not come from physical capacity,
It comes from an indomitable will.

– Mahatma Gandhi

CHAPTER 4

HOW DID I GET INTO SOLAR?

In the contradictory energy scenario of today's world, could the use of Solar energy fulfill all our energy needs be a solution? Possibly, yes! I can say this from my experience, as I have been working in the Solar energy field for over 20 years now.

I have been doing lots of experiments in the domain of Solar energy, with the best scientists in the world on one hand and with hardly educated people (in rural areas) on the other hand. On one side, I have been publishing scientific papers in international journals, whereas on the other side, I am writing training manuals for villagers. I have been working with policy-makers as well as practitioners of the Solar field. Not only this, I have the experience of being a Solar entrepreneur, though not a very successful one!

Now, the question which follows me, *"How did I get into the Solar field?"*

Well, it has an interesting beginning, something which I could never forget. After dedicating 20 years of my life to SOLAR and reaching this juncture, when many people, friends, family members, media refers to me as "Solar Man," I feel humbled.

Overall my journey has been very fulfilling.

Let me share with you, how it all happened.

It started with an address of a motivational speaker in my Engineering College in the year 1996, in Indore, who said, *"Engineers should not take up the jobs, rather they should create the jobs for countrymen."*

What would a 20-year old Chetan have understood from these words?

But somehow these words struck me, and made a lot of sense to young Chetan, underlined by the scenario that getting a job was an issue even in those days.

I don't recollect exactly why, but an idea came to my mind for establishing a factory for producing plastic footwear (slippers and shoes). At that time most of the people in my area used to wear rubber slippers, plastic shoes and slippers.

My village, Nemit, is a small village of about 800 people, still having almost the same population, in fact, the Government should award my village for maintaining the same population. There is no bus connection to my village, even today. At that time no one owned any vehicle, including my family. I also don't remember anyone using a bicycle, we either walked or used a bullock cart to go to nearby villages.

I used to enjoy bullock cart rides though there were no roads. In the rainy season, I remember that we had to walk through knee-deep mud to move in and move out of my village. When I used to go out of the village, I remember walking behind my father, in the narrow *pagdandis*, passing through the lush green fields of cotton, maize, wheat and rice, which were the major crops grown.

Even today, there is no post office in the village. The address of my village reads, District: Khargone, Tehsil: Zhirnya, Post: Ighriya, Village: Nemit, Madhya Pradesh.

My Higher Secondary pass father was the most educated person in the village. Going by this, I was the most educated one who was to become the first engineer, not only in my village but in the entire region.

Coming back to the entrepreneurial bug which had bitten me, a factory of plastic footwear made enormous sense to me. Why? It had a basis. As a child, we used to enjoy kulfi which was sold by the people on their bicycles, either in exchange for money or discarded material like a piece of iron or plastic footwear.

Alongside my Engineering, I was also pursuing Industrial Management Course, wherein I studied about the recycling of materials. Only meritorious students got this opportunity to pursue the additional course. As I performed well in earlier semesters, I got this opportunity.

So, the Kulfi sellers from my village helped me in firming this idea. I thought that discarded plastic footwear that they collect, can be collected from other villages as well for recycling and making new plastic footwear, and finally selling in the market. If I establish this factory near my village, many people will get employment. On a conceptual level, this looked like a great idea. It was like setting up a complete ecosystem of local production, local recycling, local utilization and local jobs.

The seeds were sown in my mind for Swaraj...

...Probably!

Once the idea of establishing a plastic footwear factory was well-placed in my mind, next was the time for execution which involved actual resources, like finances and materials. Now, it was time to give my dream a concrete shape.

I started exploring possibilities of setting up a plastic footwear manufacturing factory. Indore was a big city and a hub for economic activities. I figured out that there are several factories in Indore which made plastic footwear (shoes and slippers). I visited a couple of them, saw the running machines producing the chappals, enquired about the material requirements, the machinery required, the manpower required and most importantly, the finances required.

The smallest factory that could be established required about INR 1. 5 to 2 Lakhs (approx. USD 2500) at that time, in 1996. The entire fee of my engineering was INR 5000 (approx. USD 70), including tuition and examination fee. This was a very big amount for me and my family. Taking a loan was the only possibility. I started working in that direction, went to the bank, enquired about the loan process, got some forms and filled them up. It was kind of a business proposal.

The idea of starting a factory was so strong that I did not want to wait for my studies to get over. In fact, I planned to leave Engineering studies in third-year itself, the year in which all these things were happening.

WHEN I WAS BEATEN BLUE!

Now, the loan documents required signatures from my father as it required mortgaging a capital asset, in this case: Land.

I gathered courage and took those documents to my father, who had some inkling of what I was thinking.

The moment was like *Salim bringing Anarkali to Zille-Subhani!*

When I showed the documents to my father, all hell broke loose!

He thundered, *"You want me to sign these documents, mortgage my land, how dare you think I would do that, mortgage my land for your joota, chappal business?"*

He said, *"On top of it, you want to leave your engineering mid-way for this, chappal making? How can you even think of that?"*

I argued, *"This is a brilliant idea and I have done so much preparation for this, where I will generate employment in my village, and there would be no need to look for a job."*

Till now he was trying to convince me that I should finish my engineering and then think about all this. But his advice fell on my deaf ears, I was persistent. I was in no mood to give up.

Now, the real action began, out of the blue he took out his stick and literally beaten me blue. The scene was just out of the Bollywood movie, I had bruises all over my back and I was crying like a baby. It was not only painful physically, but mentally too, as parents did not understand which I thought to be a wonderful project. I cried and cried, the idea of setting up a plastic shoe and slippers manufacturing unit in the village got flooded with tears.

Finally, I was back to my engineering degree. The beating was so strong that for the next few days I kept on crying. I had to give a presentation in a seminar as a part of the course, and I broke down during the seminar.

When the teacher asked, *"Why are you crying"?*

I could not answer. Now I look back at those times with nostalgia!

TIME TO PURSUE HIGHER STUDIES

I completed my Engineering, got a job. But my very well-respected teacher and the Head of the Department Prof. Prakash Vyavahare told us that we should go for higher studies as specialization will help us in future. Many of my close friends chose to study further. I did well in the entrance examination and got the admission for pursuing a Master's degree in Microelectronics in IIT Bombay.

This was the next phase of my life…

…Higher education started with the best minds in the field, the best professors, who impressed me not only with their knowledge in the field of microelectronics but also with their simplicity.

However, deep inside, the idea that I should do something for my village stayed with me. Soon I would find like-minded people. There were some social activities happening in IIT. I got in touch with organizations like Pratham which worked for school education and CRY, which worked with underprivileged children etc. One such group in IIT was the Group for Rural Activities, where Vasu became my good friend.

He proposed, *"Chetan, why don't we visit Anna Hazare, who is doing remarkable work in his village - Ralegaon Siddhi?"*

I joyfully replied, *"Why not?"*

Meeting Anna Hazare had a great impact on my life, not for the work he was doing but the advice that he gave.

When I told him, *"Annaji like you, I am also not planning to get married and want to dedicate my life to social work."*

He advised, *"You must get married, and if you do social work then, it will be much more effective, otherwise people will say that as you are unmarried, you are doing social work."*

The advice was logical, which helped me in changing my mind. After coming back to IIT, I shared this with my friends that I wanted to get married now, I became a laughing stock for quite some time. But it was a wonderful news for my parents!

TIME FOR JOINING THE JOB

The time just flew in IIT, now I was about to finish my masters. By that time, I had also secured a very well-paying job in a multinational company, Texas Instruments. This made my friends and family members very happy.

Somehow, I was not happy as I wanted to get into the social field. I was thinking if I design Integrated Circuits or ICs, the chips that are inside our TVs and computers, who will benefit from it? Certainly not villagers! The benefit of all these technologies reaches villages after a long gap. So, I genuinely looked for opportunities in the social sector.

One of the ideas was to do Ph.D. in social sciences from Tata Institute of Social Sciences, I went to meet a couple of faculty members there, but they did not entertain me as I had a different background. Then I went to an NGO, Development Alternatives, for a job, but I was denied, citing the reason that I was overqualified.

When I discussed this with my guide Prof. Rakesh Lal, he suggested, *"Why don't you do Ph.D. in Solar Energy? This way you can pursue higher studies and with Solar energy technology, you can help the rural communities as well."*

I liked the idea.

I started reading about Solar cells in the library, a device that converts sunlight into electricity. I referred some books for a basic level of understanding and read some research papers for understanding the current status of research in this field. What I understood from my basic reading was that Solar cells are being developed all over the world but the efficiency of Solar cells is not high, besides prohibitive costs. There was a need to address this. With this basic knowledge, I made several applications for Ph.D. abroad, hoping that I will get a scholarship. I tried to approach some faculty members within IITB also, but was discouraged to pursue Ph.D. within IIT.

While I was trying all this, it was time to bid goodbye to IITB. The job was in hand, so there was no worry. I went to Bangalore and joined the job of application-specific Integrated Circuit designer in Texas Instruments. Now, I started earning money, a good amount of money. The company provided all the basic facilities, sports facilities and flexible work culture. It had an awesome building like that of a five-star hotel. Soon I started taking part in all kinds of activities, performed well in all aspects and earned the appreciation of management.

Parents were happy now as I was doing well in a well-paying job. And of course, there was a pressure to get married. I always wanted to marry someone who can take care of my village life as well as stay in the city with me. I thought only a girl living in a village can balance this life. With these conditions, parents selected a girl, Rajni, who belonged to a farming family. I couldn't have been happier with my parent's selection. Rajni, not only balanced the life well, that of a village and a metro but also became my biggest support system for all my decisions which shaped my life going forward.

A TRULY SPECIAL ENGAGEMENT GIFT

When I went to the village for engagement, someone came to me with an envelope in his hand and said, *"There is a special gift for you."*

When I opened the envelope, my eyes moistened, I was amazed at this development in my life. I could not believe what

I was holding in my hands? It was nothing but scholarship letter from Interuniversity Microelectronics Center (IMEC) from Belgium for Ph.D.

Not only they offered me a Ph.D. but also the topic of my choice which was "Development of low-cost, high-efficiency thin-film Silicon Solar cells". How amazing! The topic of choice with a scholarship. I couldn't have asked for more. I did not have a second thought in my mind regarding quitting the job, however, everyone else resisted.

My MD told me, *"We will pay you much more if you stay back in the company."*

Many people discouraged me saying, *"After Ph.D. you would just become a professor at some University."*

Also, at that time, in the year 1999, Solar technology was not much known in the country. The future did not look very bright after Ph.D. but deep inside, I was convinced that I need to take this opportunity.

I realised the need for learning more about Solar energy and Solar cell technology, so that I am able to help rural communities in some way. While taking such decisions, my village background was always very useful, my desire to help the village communities made the decision-making very easy for me. Without thinking twice, I resigned from the well-paying and well-respected job to pursue my dream of working for my village and thousands of such villages in India.

This is how I got into the Solar field.

CHAPTER 5

WHEN I DIVED DEEP INTO SOLAR TECHNOLOGIES

Going to Europe for my Ph.D. was a very exceptional instance in my life, not only for me but also for my family, relatives and friends. Though I was overwhelmed with the opportunity, I remained humble. I never boasted about this great happening in my life.

Finally, the day arrived, when I set out for Belgium, which was set to change the course of my life. It made "Solar" one the most important mainstays of my life.

I came to know about Arun, through my company, who was also going to join the same institute for the Ph.D in a different area. We got in touch and landed in Belgium together.

Dr. Renat Bilyalov received us at the airport. As soon as we were out of the airport, the car zipped in at the speed of 130-140 kmph, this was unimaginable in India, it was a thrill to watch the speedometer.

We reached the institute and finished the joining formalities in a couple of hours. The institute was outside the city, he just showed us the direction of the city. Both of us started walking on a road, where nobody else was walking, only cars were passing by. It was only sometime later we realized that it was a highway and nobody was allowed to walk on it.

Well, this is how my search for research on Solar cells started.

DREAM COME TRUE MOMENT!

I eagerly looked forward to the next day, when I was to join the Institution for my research on Solar cells. I excitedly reached IMEC, it was a "dream come true" moment. IMEC is a very famous institute in the world for its facilities particularly for conducting microelectronics and nano-electronics based research. The best companies in the field were collaborating with IMEC to conduct their research, this way significant funding came through from these companies.

After joining, I became part of the Solar cell group in IMEC which constituted 64 people, according to me it was quite a big group.

I wondered, *"64 people doing research on Solar cells?"*

In India, I heard of only a few people working on one topic, and here, so many people were just doing Solar cell research.

In the first few days, I got introduced to my group members followed by an introduction to the laboratory. After which I was introduced to the topic that I was going to work on, and within seven days I was in the laboratory starting my first ever experiment on Solar cells.

How fast and swift it all was!

The laboratory of IMEC was one of the best in the world. In technical terms, the space used for research lab had to be 'ultra-clean', in the scientific sense, as the dimensions in the microelectronics circuits are in the order of a millionth of a cm and even the tiniest dust particles, not even visible to the naked eye can damage the circuit. The cost of operation of such a lab is very high.

I was stunned to hear that the cost of one day of usage by one student was about USD 150-200. Also, some of the highly toxic gases were used in such research, which needed a very elaborate mechanism to detect any leaks, hence appropriate protection was needed. For entering the lab, we were covered from top to bottom; there was a cover for the feet, the body, the hands, the head, the eyes and the nose so that the laboratory environment does not get contaminated. Within 8-10 days of joining, I started doing my Solar Cell related experiments on a piece of semiconductor called Silicon.

MY FAVOURITE WAFER

Let me tell you about Silicon which is by far one of the most amazing materials having an enormous impact on our lives and on the sustainability of human lives on the planet. This is a semiconductor used for making all types of electronic circuits or electronic chips. So, our computers, mobiles, calculators, cars, and every other electronic device uses this material. Interestingly, the same Silicon is also used for making Solar cells. This way, I have had a very close relationship with Silicon.

Also, silicon is required to be produced with high purity standards. The purity we normally hear of 99.9% or 99.99%. The Silicon is normally required to be produced in the purity of 99.9999999% for microelectronics and Solar cells applications.

With so many applications, who would not fall in love with this material? Today Silicon is the hero of the Solar cell field. More than 90% of Solar modules in the world are produced using Silicon. For Solar cell research I used to start with a very thin piece of Silicon, typically referred to as Silicon Wafer, which looked almost like a mirror.

Soon after getting access to the laboratory, I started doing my experiments. My supervisor asked me to perform some experiments with Silicon wafer. There were 36 sets of experiments to be performed with a certain set of experimental parameters. I had some thoughts in my mind regarding the design of experiments involving some new experimental parameters, which my guide did not approve.

It was late in the evening, at about 8 pm, I was still performing experiments. I performed 35 experiments as per the plan, for the last experiment, I could not hold my thoughts and tried the experimental parameters that were not part of the plan. And, the experiment failed miserably, it appeared that the Silicon wafer broke in thin layers, and pieces of thin layers of Silicon wafer started coming out of the experimental setup and started floating in the chemicals. I was scared!

Firstly, because I did not follow the plan and secondly, the experiment resulted in quite a mess in the experimental setup.

WHEN MY ACTION DID NOT HAVE AN EQUAL AND OPPOSITE REACTION!

I closed the experiment, painstakingly cleaned the mess and left. I was ready for shouting from the supervisor the next morning. I described to him what happened in the last experiment.

He said, *"I want to see the thin layers that came off the Silicon wafer."*

After looking at those thin layers in the lab, he said *"Wow Chetan! this is wonderful."*

I was surprised and could not believe my fate as I was expecting an exact opposite reaction. We discussed in more detail and realized there was great potential in the failed experiment also. It was the beginning of the new chapter in that particular field of experiments, which is now technically known as Porous Silicon lift-off technique using electrochemistry.

I experimented more often using accidentally discovered new set of parameters, mastered the technique and obtained three US patents, related to the technique. This became the main focus of my doctoral thesis. I also published several research papers on the technique which is now one among the known techniques for making thin films of monocrystalline Silicon.

Since with this technique, we could make the Solar cell of high quality but by consuming less material, there was a great potential for making low-cost and efficient Solar cells. This is exactly the reason why I was interested in the Ph.D. and this was the purpose for which I got the scholarship.

Well, it is said that most discoveries happen accidentally and, in my case, it was no different. After I graduated and obtained my degree, there were two companies that I knew were trying to commercialise this particular way of making thin-film high-efficiency Solar cells. Later I visited one of these companies in the US, where I was happy to see the big machines using the techniques that I developed during my Ph.D.

RESEARCH HAS ITS OWN LIMITATIONS

Unfortunately, this particular technology could not become very successful as there was an inherent limitation of making these Silicon thin-films and the complexities of handling the films. There was a need to develop high efficiency and low-cost machines. I realized later that only 5% of all the research performed in the world eventually gets utilized. Many times, the new technologies face limitation due to non-availability of funding, the initial high-cost of developing machinery, and by the time one perfects a technology, some other new technology makes it obsolete.

My technology could not come in the mainstream production of Solar cells, otherwise, I would have been a millionaire with my patents. My research during Ph.D. was a great learning experience. I understood Silicon very closely, its production, its processing and turning it into a useful device like Solar cells.

WHY DEVELOPING WORLD DOES NOT TAKE - OFF TECHNOLOGICALLY?

Handling of very high purity Silicon wafers, the clean environment that requires to process them and sophisticated arrangement required to maintain the laboratory, makes Solar Research and Development unaffordable in many countries of the world.

I was surprised to know that the cost of my usage of the laboratory was about USD 150 per day, now I realised why there are very few good laboratories of Silicon processing, even in developed countries and much less in developing countries.

Anyway, with my rich experimental experience during my Ph.D. in Solar cells, I could clearly see what was lacking in India, why world-class Solar cell research is not happening there and what it would take to set up such facilities. I also learned to appreciate the development of the entire cycle of any technology from beginning to end and its requirement to be a successful technology, that can be commercialised.

Here, I also got the opportunity to attend conferences and visit some laboratories not only in Europe but also in the US and Japan. During my third year of Ph.D. I visited Tanzania in

Africa for 12 days for a teaching assignment in a workshop. I felt that the audience enjoyed and understood my explanations. All this gave me further exposure, enhanced my understanding of the possibilities of Solar technologies. I realised that a wide range of work with Solar technology was possible and that it possesses immense potential. In a way, every year of my Ph.D. was an enriching experience in Belgium!

SEEDS OF BECOMING AN AUTHOR GERMINATED!

To keep myself updated with the latest technological developments in the field, I started reading research papers and books from other laboratories around the world. Somehow, in this process, I developed a desire to write books which could explain the subject in a simple way, the science and technology behind it and their applications.

Somehow, this desire could not take wings then but resurfaced after my joining IIT Bombay. Till date, I have written several books, each one of these widely utilized, a couple of them are being used as textbooks in many universities.

"Permanent head Damage" and "Pretty heavily Depressed" and many such other full forms are being used in academics to describe the state of Ph.D. students. There is no fixed curriculum that you finish and graduate. So, it happens very often, at least once during the doctorate studies that the student gets frustrated and decides to leave. It happened with me as well, somewhere near the end of the third year when I decided to leave and started looking for a job but soon better sense prevailed, and I crossed the finishing line.

The experience gave me a strong foundation, not only it provided me in-depth knowledge of Solar energy but also helped me in gaining width of Solar energy field, both of which were essential for drawing a complete picture of the world. Indeed, this platform played a great role in my life, and in some way, in the lives of many other people.

CHAPTER 6

WHEN LIFE SUPPORTED MY PREFERENCE: TO BE BACK IN INDIA!

Those were the times, the late nineties when everyone wanted to go to Europe or the US, with the aim of settling down there. Once there, they never thought about returning to India. I guess this was and is still true for the citizens of many developing countries. In search of a better life, they chose to settle in developed countries.

I was already there in Europe, but this was a kind of scary thought for me. I wanted to do my Ph.D. in Europe so that I could learn Solar technology and use it for the betterment of society in India. My aim was to come back to India and not to settle there in Europe. Even the thought of settling in any western country disturbed me.

Now, it was time to work out a methodology which enabled me working towards my plan of coming back to India. I started with fixing a date five years advance and started working backwards. After a few deliberations, I fixed my birthday, nearly five years in advance to go back to my country, that is on 21st May 2004.

In order to ensure that I go back, I started regularly keeping a count of the days. Every day on the top of the diary page, I would write the number of days remaining, i.e. 1225 days to go, 1224 day to go, so on and so forth. This way every single day was a reminder for my plan.

While I was in Europe, not only I got enriched with the knowledge of Solar Technology but also my life got enriched in many other ways. I used this time to do many other meaningful things.

Now the marriage beckons!

MY IDEA OF A SIMPLE MARRIAGE

One of the significant things which happened in the first year of my Ph.D. was my marriage. It was a nice thought to marry someone from a village who should join me in Europe to experience life there. Besides, I also wanted a simple marriage. Hence, I decided to opt for a group marriage ceremony to keep the expense minimum. In India, marriage is a very big occasion, so much so that families borrow money to have a good marriage function, beyond their capacities and eventually suffer a lot in repaying the debt.

This was with the purpose that if I tell someone to have a simple marriage, then first I should also do it. My idea was not liked by our families (mine & bride's), as they wanted a good marriage particularly when the boy was pursuing Ph.D. in Europe. It was very hard for me to convince both families on this. Somehow with the help and guidance from friends, I stuck to my decision and finally got married in a group marriage ceremony.

Later my wife Rajni joined me in Belgium.

I had an interest in photography, so when I got my first scholarship, I bought a professional camera and started learning photography. So much so, I went to Washington DC to participate in a photography competition. One can imagine what it would have been like spending so much money particularly when you are a student? It was a very difficult decision, but I was mad about photography then.

When I presented my photograph to the jury members, they liked the concept of my photograph very much. Amidst 700 photographers who had come from all over the world for the competition, I got the second prize of USD 2000, which not only covered my expenses but left me with some savings also. I also participated in several exhibitions in the town and received really good remarks for my photographs.

Now another diversified interest, I learnt to play flute and *tabla* as well. A fellow from Bangladesh taught me *tabla* and the person from Maharashtra taught me playing the flute. In general, I am a good learner so in a short time, I could learn both but could not master either of them. I just managed to play. I also performed in few programs.

LOVE FOR LANGUAGES & SPORTS!

I also enjoyed learning various languages. I learnt quite a bit of Bengali and Flemish/Dutch. After quite some efforts I could read, write and speak both these languages reasonably well. I also gave a shot to learn Spanish.

In terms of sports, I used to regularly play badminton, but as an Indian, I loved cricket over any other sport. We started playing cricket in the open areas but in Europe's weather, it is not possible to play throughout the year. So, we started playing indoor cricket, probably for the first time in the history of the town, indoor cricket was being played. Many times, I used to be the captain.

One of the most unusual things that I have done while pursuing Ph.D. was learning *Bharatanatyam*, which was a strange but interesting setup. We were learning *Bharatnatyam*, an Indian dance form in Belgium from a Russian teacher. What a unique combination it was!

Not only me, but my wife also joined me in learning *Bharatanatyam*, both of us learnt a very basic level, in fact, performed on the stage in a cultural evening in the town. What a memory it is now! How could we do that!! To be dressed in *Bharatanatyam* attire with proper makeup, both husband and wife performing together.

A beautiful memory that I will always stay with us.

WHEN CHETAN BECAME DR. CHETAN, LIFE TURNED A NEW LEAF!

By the end of my Ph.D. I had some job opportunities in Japan and Europe, but I always wanted to come back to India, so I was also applying for jobs in India. IIT Bombay was one of my favourite destinations and it so happened that I was offered the post of Asstt. Professor in IIT Bombay, just before finishing my Ph.D.

While all this was happening in my life, my clock was ticking (of returning to India), the number of days came down to hundreds and eventually to double digits. Once, when I was very close to my Ph.D. defence examination, in the stress of preparation, I stopped putting the count in my diary. Somehow, I completed my work and got clearance to do my Ph.D. defence in the month of May 2004. It was successfully defended and from Chetan, I became Dr. Chetan.

Now, it was time to return to India from Belgium. Finally, the day came when we took a flight to India. The day was 20th

May 2004 and we landed up in Bombay on 21ˢᵗ May in 2004 at 2 am. The date which I had set nearly five years in advance.

How strong and powerful these coincidences were? It is very difficult to describe these happenings. I am sure within our human capability it is difficult to plan things in such a perfect manner. There have to be other forces which help us in doing, what we are doing. Such things tell you that you are not the doer, you are merely a medium of happenings around you.

I learnt during my job in Texas Instrument that *"life tends to support our preferences,"* I strongly believe in this. I made this line as my Email signature and very often I referred this line to people. The fact that I will end up exactly on 21ˢᵗ May which was planned 5 years ago is an indication that the situations came together in such a manner that I could get what I preferred i.e. coming back to India.

LIFE'S SOME INVALUABLE LESSONS LEARNT!

All of us should believe that actually life tends to support our preferences if you wish to become something or if you wish to achieve something. If you really desire something from the bottom of your heart and if you take sincere actions, the world will conspire to make sure that you achieve what you want, be it good or bad.

I have experienced this several times in my life, that I have been able to achieve whatever I have aspired for. My belief has become stronger that life tends to support our preferences.

In fact, there is a very famous dialogue from Bollywood movie where famous Bollywood actor Shahrukh Khan says, *"When you want something to happen from the heart, the entire universe conspires to bring that to you."* How amazing? Isn't it?

All these learnings became the foundation stone of my life which have shaped today's Chetan. These learnings are the base of my thought process for positivity and activities.

These learnings are the base of my actions in the field of Solar energy, these learnings made my soul which is constantly thinking of sustainability of planet Earth and that of the human race if I can say so!

PART - C

PUTTING SOLAR KNOWLEDGE TO PRACTICE – LEARNINGS ON SOLAR ENERGY

Live as if you were to die tomorrow,
learn as if you were to live forever.

– Mahatma Gandhi

CHAPTER 7

THE JEWEL IN THE CROWN: 100% SOLAR EDUCATION PARK

There is a difference in learning theoretically and practically, but both are important. Mahatma Gandhi used to say that training of three Hs: Head, Heart and Hand is important for balanced growth. After spending 20 years in the Solar field, I can confidently say that it is the training of Heart (through Art of Living) and Hand (through practical work in the field), more than the training of Head (through education for degrees) that gave me clarity on the what needs to be done.

Unfortunately, many people do not get training of all three Hs, hence there is an imbalance in vision, in setting up targets and means of achieving it. The same is true for societies and nations.

Having studied in a single classroom primary school in my village in MP, and getting the opportunity to pursue doctorate in the most prestigious University of Europe, was a matter of getting the opportunity. I have always believed that many children with similar opportunities would be able to achieve the same feat. Unfortunately, that doesn't happen often. Therefore, I always wanted to do something for the village children and one of the ideas was to provide them with a good education.

Earlier, I had worked with several NGOs for social causes. There I experienced that help provided to the students was insufficient and of low quality, thus failed to bring required change in their lives.

Though any kind of support is good but probably most of the time, it is not the life-changing support. One of my other initiatives was in the form of ROSE (Rose, an Organization for Supporting Education), an NGO, which I started in Belgium. This too did not give us the desired result.

I always compare these experience with the electronic device called Diode which requires threshold voltage to operate. I guess engineers would understand this. Below the threshold voltage, the Diode does not switch-on, irrespective of how long you provide the voltage. But as soon as you provide the voltage above the threshold level, the Diode switches-on.

I always felt that the support to the deserving candidates should be above the threshold level so that they can switch-on in their lives and are able to stand on their own. Hence, I wanted to develop an educational campus which is as good as any other good school of the state or even country. At the same time, the campus should provide affordable education to the students. The past experience and this thought process gave birth to Education Park.

The Education Park is a non-governmental and non-profit organization located in Bhikangaon, Khargone district of Madhya Pradesh. This park is serving the educational needs of 40-50 villages located within a radius of 15-20 km. The objective of this Education Park is to "provide high quality and affordable education and training in rural India" by focusing on 'Education', 'Economy', and 'Energy' (the 3Es). Built on the Solar-passive architecture, Education Park is a case study in itself.

ENERGY ASPECT: THE HIGHLIGHT OF THE EDUCATION PARK

How can we make a school campus that runs fully on Solar Energy?

This was done during 2009-10 when Solar energy was very expensive.

The simple way to achieve this was to use the formula, *"Energy saved is energy generated."*

In the region where the Education Park is located, in 2009-10, the power cut was a norm which varied between 10 to 20 hours per day. In this scenario, ensuring continuous power supply to Education Park was not possible until an alternative arrangement like a diesel generator and/or inverter was deployed. These solutions were expensive considering their operational cost; hence it was decided to use Solar power for the daily operation of the Education Park. The work began, by taking the first step of keeping daily energy usage to a bare minimum.

If the concept of 'Solar-Passive Architecture' is followed at the designing stage itself, the energy consumption of the building reduces significantly. This way up to 80% of energy requirements can be reduced, particularly for buildings which are used only during the daytime, such as school buildings. Mr. Jiten Prajapati was the main architect, supported by Mr. Malak Singh, for the Solar-passive architecture.

For this, an integrated design process was followed, which took into consideration, the climate, the social and environmental context, and the functional requirements of a school that enabled a comfortable and sustainable environment in the premises.

SOLAR - PASSIVE ARCHITECTURE FOR EDUCATION PARK

Since the school is located in a remote area with scarce resources, other aspects, such as water and waste management, on-site power generation, promotion of pollution-free environment, and low-cost construction technologies have also been incorporated in the design. The attempt was to provide a centre for learning that is appropriate in the local context and provides a stimulating environment for the students at large.

The climate in Khandwa– Khargone district of Madhya Pradesh in India is composite in nature which shows three distinct climatic conditions throughout the year. Therefore, the school is designed for a composite climate.

Resisting heat gain, particularly during summer, is the major design consideration. During monsoon, ventilation helps to alleviate discomfort. The basic design strategy is inspired by the concept of a courtyard, which would act as an environment modulator for the spaces around it.

For better thermal performance, many features have been incorporated; these include, doubly loaded green corridor with a North-South orientation, internal courtyards to provide comfort, and semi-open interaction spaces. Windows are appropriately shaded on both sides, North and South, with deep *chajjas*, (cover of a roof) on the North and South.

The geometric form of the building has been developed in response to Solar geometry. Walls of the building are aligned in East-West direction, to minimize the heat gain from the sides. Heat gain from the top is reduced with the help of filler slab. And in order to make better ventilation, windows were made perpendicular to the natural wind direction of summer, which is the North-West direction, this way triangular-shaped windows were designed.

This use of filler slab roof and the *chajjas* on the sides reduced significant heat gain inside the classrooms. As a result, a temperature difference of 11 degrees centigrade was measured between outside and inside the classrooms in the peak of summers, that too without using any fans.

Thousands of trees have been planted on-site to rejuvenate the environment and to pre-cool the air entering the building premises. Low-cost technologies such as filler slab, un-plastered walls, RCC door frames, composite RCC, and load-bearing structure have also been incorporated in the design and architecture.

The simulation study was carried out to investigate the performance of two different functions of windows: (a) daylight and (b) vision window. It is found that the side lighting with daylight windows on both sides and the vision windows on one side provides adequate sunlight in the classrooms.

To optimize the daylighting levels in the classroom on the southern side, a light shelf is used. In general, high surface reflectance of walls, light shelves and ceilings, and a higher visible transmittance of glass improves the performance of daylighting.

Various green building methods, for example, using renewable energy sources, water and waste management, energy-efficient fixtures and equipment have made the campus, self-sustainable and ultimately a zero-energy campus. Having designed and fabricated such building, the next step was to test the performance.

All these aspects of Solar-passive architecture are very important if one needs to run the premises fully on Solar energy as it helps in reducing energy consumption significantly, which in turn helps to reduce the cost of Solar installation.

The performance of the school building in terms of comfort has been assessed by taking onsite measurements of daylighting, temperature, and humidity. A post-occupancy evaluation of the comfort level has been also carried out.

There is a high-level of satisfaction among the students. In fact, daylight is found to be sufficient and no tube lights are required during the school hours. No fans have been installed in the classrooms, as adequate cross-ventilation is available. However, in recent times, when the electricity supply has improved in the region, people got the habit of using fans even in winters, as a result, the students demand fans in some months.

The energy needs of the Education Park are being minimized by using Solar-passive architecture elements, while the remaining energy requirements of electronic loads (computer and printer etc.) and lights and fans are fulfilled by a Solar PV system.

THE CAMPUS RUNS ON 100% SOLAR ENERGY

In order to supply electrical load in the Education Park, a standalone PV system has been designed and installed. The entire campus since its inception in 2010, always runs 100% Solar energy. No electricity connection has ever been taken to the campus; this was only possible as we had cut down 80% needs of energy through the use of Solar-passive architecture. Many people visit the campus and admire its architecture and Solar solutions. I take a lot of pride in establishing a zero-energy school campus.

I acknowledge the contribution of all those noble souls who have contributed both in terms of money as well as efforts to make such a beautiful campus and sustain it.

If we can make one campus like this, why can't every campus in the country and in the world be like this, particularly, the academic institutions, which are the temple of knowledge? It is important that the knowledge of sustainability is ingrained in the young minds, both in theory and in practice. The park is expected to have a multiplying replication effect in the country.

The need of the hour is to involve the students (youngsters) and teachers alike in this task of monitoring the energy parameters on a regular basis. Facilities like these may ultimately be recognized as a classic example of capacity building initiatives. After all, energy efficiency measures should go hand-in-hand with the actual deployment of renewable energy like Solar energy in case of Education Park.

It has taught me many lessons but, first and most important was how can we avoid using energy before even thinking of generating energy. How it makes Solar solutions viable? I believe, this lesson is a key aspect of sustainability and a very important aspect when one needs to switch to Solar completely.

CHAPTER 8

MY EXPERIMENTS WITH SOLAR TRUTH

After spending nearly a couple of years in IIT Bombay (IITB), in 2006-08, I thought of learning about making of Solar lamps.

Prof. Chetan thought, *"What would a person with a Ph.D. in Solar technology need to learn about making a Solar lamp? It should be a cakewalk, isn't it?"* But that was not the case at all.

For villages where electricity supply was erratic, where there was no electricity or where the power cut for 18 to 20 hours every day was the norm, what would be the most interesting thing a government, an NGO, a philanthropist or a highly placed official would do?

Provide Solar lamps.

WHY SOLAR LAMPS?

Solar lamps as standalone products are quickest and cheapest lighting solutions available to people.

The Central government, as well as State governments, run their own schemes to provide subsidized Solar lamps. Not only governments but the likes of Amitabh Bachchan and Sachin Tendulkar would also donate lamps, so I also decided to join the bandwagon.

"Why not me? After all, I am a Ph.D. in Solar cell technology from Europe, I knew Solar lamp better than anyone else," I told myself.

Since helping the society was one of the reasons that I wanted to come back to India after my Ph.D., after joining IIT Bombay in 2004, I started exploring options for fulfilling this objective.

While I was in Belgium, I came in touch with Mr. Bharat Jhawar, a CA by profession, a well-respected and dedicated social worker. He introduced me to an organization called *Seva Bharati*, which runs very unique schools called *Ekal Vidyalayas*, meaning single teacher schools.

As they were in remote locations, hence lacked connectivity and access to basic things. These schools are run in the evening hours, considering the fact that many kids have to work during the day time in their farms with parents or have to take care of their younger siblings at home. As *Ekal Vidyalayas* run in the evening, invariably it meant that these needed light. I visited some of these villages which were remote and lacked infrastructure, once, while approaching a village, I thought we lost the way.

I stopped and asked my companion, *"Have we lost the way, where is the road?"*

He replied, *"the place where you are standing right now is the road."*

TOUCHED BY THE PAINFUL SIGHT

When I reached the village around sunset, I was taken to a small shed, which they called school. It was time for school to start. Now, a sight deeply pained my heart: I saw children coming to the school, with a few books in one hand and a kerosene lamp in the other.

In such moments, you just want to pause, you want the world to stop as it is, you want to think and wonder why things are still like this? You would like to feel and share their pain; you would like to cry in pain.

The image of a very advanced laboratory in Belgium flashed in my mind. Aren't these two scenarios presenting a sharp contrast? As big a contrast as day and night? I myself studied in the light of kerosene lamp but that was in the late 70s, and I did not go to school carrying kerosene lamp.

Well, the pause did not happen as around 20 school children gathered with kerosene lamps, and school started with a beautiful prayer.

A little more study would reveal that not only these kids, but there were about 100 million school students within India who were suffering due to either no electricity, poor quality, or poor supply of electricity. A similar situation existed in many other parts of the world.

Around the same time, the year 2007-08, I was doing research on Solar cell technology and on highly advanced nano-materials like Silicon quantum dots for making better and cheaper Solar cells and Carbon nanotubes. Applied Material, a leading semiconductor equipment manufacturing company, sponsored one project and I got in touch with their management team. Once over dinner, I discussed with them the situation of the school I visited, and my desire to provide them with the magical Solar lamp, hoping that it would solve a big problem of the students.

Probably, my good work so far with them helped, they happily sanctioned funds for 500 lamps. It was a big number, I jumped with joy! Now kerosene lamps would be replaced by Solar lamps.

I discussed the plan with Mr. Bharat Jhawar of providing one Solar lamp to each *Ekal Vidyalaya*, which meant covering 500 schools, implying reaching out to 500 villages. I knew only a few schools; therefore, the identification of schools and villages was done by Mr. Jhawar. Logistically, sending Solar lamps to 500 villages, in different states of India was an uphill task. Somehow, Mr. Jhawar managed it well.

I remember that all the 500 lamps were delivered at Keshav Dham, a residential ashram for differently-abled students in Khandwa, Madhya Pradesh. We charged these 500 lamps in the sunlight, kept all the lamps in a room and switched them on after charging. What a bright light came out from all the lamps? It was almost like that another Sun had risen, a baby of the father Sun, and this baby Sun was going to remove darkness from the lives of all those students who will study in its light.

MY LEARNING OF SOLAR LAMP BEGAN

All the 500 lamps got distributed in 500 villages amidst fanfare, the idea was to give one Solar lamp in each *Ekal Vidyalaya*. All students in the school will sit around the lamp and study. But who knew at that time, where it will lead to? In no time, issues started cropping up. Some would say light is not enough, in some cases lamp would not run for sufficient hours.

I could not do much as lamps were already distributed. Scientist in me started to optimize the usage. There was a need to raise the height of the Solar lamp so that the light will spread and student can encircle the lamp and study around it.

This was a CFL (Compact Fluorescent Lamp) based Solar lamp with heavy lead-acid battery and wide base. As a result of the wide base, when we increase the height of the Solar lamp, the base of the lamp itself cause shadow beneath it and make the area unusable. Then an idea clicked, why the CFL, the light source has to be attached to the base housing battery, we can detach it and hang it separately.

With little effort, I trained a few to separate the two. Now, one can hang the CFL on the height, facing downward and keep the base of the lamp with battery separately. This greatly solved the shadow problem, but ultimately it was only one lamp among 20-30 students. I felt it is not going to be sufficient. However, the villagers were very happy with this one lamp itself. For them, it was a big improvement, a big incentive, it not only saved the inconvenience of carrying a kerosene lamp but also the money spent on kerosene.

After some six months, more troubles started surfacing: Solar lamps will not charge, the battery did not work, it will just run for 1-hour maximum, the panel stopped functioning, the fuse got blown up and many more. The lamps were distributed in 500 different locations in different states of India. Now the question was how to repair these lamps? How to provide solutions to the issues cropping up? I had no idea. I tried giving them some solutions, but deep in my heart, I knew this is not going to work.

My palace of cards started crashing, faster than I could ever imagine.

So far, I only knew how to design a Solar lamp technically, but in the absence of practical experience of a real-life situation, the design is not going to work properly or worse, it is going to fail. The latter was the case here. And, as far as I know, all the Solar lamp programs that were happening or happened, by the governments, by NGOs, by the Philanthropists, eventually met the same fate.

TERI's LABL REJUVENATED THE SOLAR PROGRAM

About a year later, I got in touch with another program. It was a world-famous Solar lamp program that was being run by The Energy Research Institute or famously known as TERI. The program was called Light a Billion Lives (LABL), which aimed to touch the lives of a billion people. Who wouldn't be impressed by its objectives? I was one of them.

The model of LABL was quite different. It had a centralized charging station, wherein all the lamps used in a village gets charged at one central location. People were not the owner but just the user of the lamp. They had to pay money for daily use of the Solar lamp, about Rs. 2 - 3 (3-5 USD cents) per day. In this mode of operation, people were required to visit the charging station twice a day, in the morning to keep their Solar lamps for charging and in the evening to collect it. A beautiful program by design.

On my request, and some persuasion, I got the LABL Solar lamp program to be implemented in three villages. One in my own beautiful small village, Nemit and two in other neighbouring villages. And, history was repeated!

The beginning was great, people will pay the money and take the lamp. It was well reported in the media also. But in a month's time, when the honeymoon period was over, the problems started surfacing. People sometimes will not pay money, sometimes they would find it hard to make two trips to the centre every single day, but worst of all, the lamp will not work for the promised time of 4 hours per night, hence customer dissatisfaction, as they were paying for it.

Some lamps failed and required repair and maintenance. I thought the repair of the Solar lamps is well-taken care off, as while installing the Solar lamps in the village there was 17-page contract document, yes 17-page document, that I signed with Solar lamp supplier along with funding agency.

I could never get a single Solar lamp supplier to repair any lamp. Several calls to supplier were fruitless. Since I was a professor at IIT Bombay, a premier reputed institute in the country, therefore, I could talk to many high-level people in the funding agency, requesting them to intervene and help, pressurise the supplier to send someone to repair the lamp. Believe me, we never ever got anyone to repair the Solar lamps.

The dissatisfaction among the users grew. More and more Solar lamps stopped functioning, and by the end of the year, the project in three villages collapsed completely.

See what had happened, I thought I knew Solar lamp, I am a Ph.D. but both these projects where I was involved, failed miserably right in front of my own eyes. I could not do much, in both cases, just watched helplessly. Same was the story everywhere else, in all Governmental and Non-Governmental projects.

This explained why Solar lamps have not replaced kerosene lamps all across. Within India itself, the kerosene subsidy provided by the government in the year 2012-13 was INR 30,156 Crore (USD 4308 Million) which in my view was sufficient to replace all kerosene lamps with Solar lamps.

These were my first experiments with SOLAR truth, which opened my eyes and paved the way for a new beginning.

CHAPTER 9

A SMALL BEGINNING OF A BIG TINY SOUL

After almost 10 years of getting into the field of Solar energy and about six years after joining IIT Bombay, I started getting some depth in Solar technology, both in terms of science and technology behind it and more importantly, in terms of applications of Solar technology.

On the science and technology side of the Solar cells, I have been writing some research proposals, getting some grants, completing small projects and writing research papers. But I was never happy with the small efforts. During my Ph.D. there were 64 people in the group.

Here in India how Chetan alone or few others can conduct research in the field that is so competitive world over, that too with limited experimental facilities?

I have been thinking of having a reasonably large Solar Cell research centre at IIT Bombay.

I have been discussing this idea with several senior faculty members, particularly with Prof. Juzer Vasi, who taught me during my masters. One of the kindest souls, he has been a great support. Prof. Vikram Dalal, from Iowa university, was the other person with whom I discussed the idea of the centre, as he has been a visiting faculty to IIT Bombay.

NATIONAL SOLAR MISSION BY GOVERNMENT OF INDIA

With luck on our side, Govt. of India, launched the National Solar Mission in 2010 with a major focus on the promotion of Solar energy. Somehow the idea of the centre clicked and got a mention in the National Solar Mission document. Then Principal Scientific Advisor to the Prime Minister, Dr. R. Chidambaram also played a crucial role here. I felt activated.

I started preparing a blueprint of such a centre with guidance from Prof. Vasi. More than 40 faculty members from 8 different departments and centres joined us by the time the final document was prepared.

It was a first-of-its-kind centre in the country. Also, the quantum of funding that we received was the first of such magnitude for Solar energy research. The Centre is known as National Center for Photovoltaic Research and Education or NCPRE.

Soon funding for the NCPRE came which filled us with enthusiasm. Prof. Vasi and I became the Principal Investigators or Heads of the National Center. We all got busy in setting up the laboratories, hiring people, taking more M.Tech. and Ph.D. students. In no time, it was one of the biggest centres in India. We started publishing many research papers, filing many patents, organizing many seminars. The national centre also got its due recognition not only in India but in some other parts of the world as well.

WRITING OF BOOKS AND TRAINING 1000 TEACHERS

After teaching Solar cell technologies for several years in IIT and recording some of the courses in the form of video, I had enough clarity, confidence and material to start writing a book, pursuing my old dream. At that time there was no comprehensive book by an Indian author, but only by foreign authors on the subject.

I decided to write a book that covered topics from basic science to applications, the fundamental aspects, technological aspects and application aspects. The book turned out to be unique, as it was written in a simple way. I used one of the summers for writing most of the book in my village where temperature was 45-46 degree centigrade.

I had to rely on battery inverter to charge my laptop. I used to sit under the trees, with cows, buffaloes and dogs giving inputs from all over! Still, I thoroughly enjoyed writing the book. This book now after 10 years of publication is used as a textbook on the subject, not only throughout the country but in many parts of the world. I keep getting appreciation from many readers, thanking me for writing such a book, for its simplicity and for its comprehensive nature. I have written several books on Solar technology with varying technical intensity. One can get all the details of the books with a google search of my name.

Education and training were one of the important aspects of NCPRE. One of the initiatives that I took under NCPRE was to teach 1000 teachers from all across the country. Prof. B. Phatak by then had pioneered large scale training program and I just used his platform.

However, I added another dimension to it. I thought only theoretical training would not be sufficient and we must provide practical training as well. How to do this, as the training of 1000 teachers was to happen through the internet in some 35 remote centres, all across the country.

I proposed the creation of Solar laboratory kits and providing these laboratory kits to all centres. There was some resistance in NCPRE but then people agreed. I took the charge and with the help of other faculty members, we developed a very low-cost laboratory kit, on which students could perform 10 experiments. The cost of the kit was just Rs. 50000 (USD 700). Many people within India and outside India were surprised that we could create such a low-cost kit.

TEACHING THE TEACHERS IN 35 REMOTE CENTRES

It was 10-days course, in the first half of the day we (Prof. B.G. Fernandes and I) would deliver lectures and in the second half the participants, nearly 1000 of them, all faculty members from various engineering institutions, would perform the experiments in all 35 remote centres, spread all across the country. I would also give them assignments and conduct online quizzes. I treated them like students. It was an amazing experience!

For the transfer of knowledge, the physical distance does not matter. The course taught me how important it is to give practical training. A possibly boring course got converted into one of the best courses conducted in a distant mode. Moreover, I learned, how easy it is to impart practical training, how the Return on Investment is very high, the benefits of imparting practical training are much more than just theoretical training.

As Gandhiji said, training of hands is very important. In general, when training by hand happens, it always becomes joyful learning.

Later I got another sponsored project where we provided the Solar laboratory kit to over 200 engineering colleges. We also wrote a laboratory manual that would go with the kit. This kit had a major impact.

KEROSENE LAMP STILL HAUNTED ME

By the year 2011-12 many people in the country started recognising me, people from academics, the industry as well as from the government. My books, my research work, the NCPRE, some awards, all these helped in holding me in good stead in the country, and earned me respect from everyone around. I guess, I became one of the recognized people in the field of Solar cell technology. I started getting invitations for talks, request for consultancy projects and for membership of committees, etc.

BUT my heart was somewhere else. I wanted to help rural communities by bringing Solar solutions to them for their immediate benefit. I kept thinking about how to solve the problems of Solar lamps. How to replace every kerosene lamp with a Solar lamp?

IDENTIFICATION OF PROBLEMS & WORKING OUT THE SOLUTIONS

I had some major learnings from my failed Solar lamp projects, the most important being that repair and maintenance were a big issue in rural areas. Firstly, in the absence of technical know-how of the Solar lamps, even a small fault was difficult to fix and Solar lamps would stop working prematurely. Secondly, the Solar lamps were not affordable by the people belonging to non-electrified or partially electrified areas, for whom lamps were actually meant for. Thirdly, Solar lamps were not available as and when required.

In my understanding of the field so far, all the three problems were the result of the manner in which Solar lamp programs were done in the country. Usually, whether government or NGOs or individuals would buy the lamps from vendors and provide to the beneficiary. In this process, though everyone had good intentions, none of them would have good program design to provide sustainable solutions. Buying the lamps and providing to communities was useful but did not make the Solar lamp affordable, hence there was a continuous dependence on funding agencies.

The technical skills were not imparted to the locals, even the most advanced programs of funding agencies would have a repair and maintenance contract with the supplier, but it never worked effectively, as a result, most of the Solar lamps would die a premature death. Since Solar lamps program were always implemented in the project mode, and for every project, one has to define the number of beneficiaries in advance, none of the funding agencies could ensure availability of Solar lamps beyond the project timeline.

Though I have described the problems in reference to Solar lamp projects. But, when I closely observed, all other Solar-based devices particularly Solar street lights face similar problems. I realized that these are the universal issues of Solar technology solutions, especially if they are meant for remote rural areas. These were the SOLAR truths, that many did not understand fully.

I felt that finding a solution to the above problems would mean, thinking of a completely new paradigm of doing things in Solar.

LOCALISATION DID THE TRICK HERE

Why can't local people make these Solar lamps? Is it a difficult technology to deal with? Would it affect the quality of the products?

The idea clicked!

However, I could see several, very strong benefits of involving local communities.

First, if the local communities make these products, they will have confidence in repairing Solar products. This will solve the problem of repairs. It would also offer an opportunity to earn money. The cost of assembly of Solar products in villages would be less than factories in a town, hence it would help in cost reduction. Since locals are involved in assembly and distribution, it is easy to keep a stock of material, this way, the problem of availability of the Solar products in local areas can be solved. Thus, localisation would address three main problems identified earlier.

SEEDS OF LOCALISATION OF SOLAR ENERGY SOWN

I called it "Localization of Solar energy". Can something as universal as Solar energy really be localised?

Solar energy, as it is not only universal but also touches each and every human being. It is a link between the universe and humans. Solar energy is universal as well as local.

To implement this idea of localisation of Solar energy, I took a sabbatical from IITB so that I could live in villages and try this model. The leave would start from 1st January 2012. Before that, I started making a presentation about the plan within IITB.

My initiative was about involving local communities in Solar lamp assembly and repairs, and therefore I needed someone to supply the components of a Solar lamp. Normally, the companies would not do this. Everyone was willing to provide Solar lamps, but not the components. During one of my visits to Vijayawada, I met Dr. B. Ranga who was running a company and providing all types of Solar solutions. I could see that Dr. Ranga was really a passionate promoter of Solar lamps. On request, he agreed to provide the components of the Solar lamps.

Now, I had to choose a location to try out the localization model. What could be a better place than Education Park? The campus which was running 100% on Solar energy and I was the chairperson of the organization, so the use of premises would be no issue.

The plan of implementation was ready, the supplier of Solar lamp component was ready and the potential users of Solar lamps were also there. Still, a few more challenges were there, like bringing the raw material to Education Park campus, communication about the initiative to potential students and cost of the Solar lamp.

SUCCESS, FAILURE, TWO SIDES OF THE SAME COIN

The raw material was supposed to come from Hyderabad, Andhra Pradesh to Itarsi in Madhya Pradesh, which was the nearest station from Education Park. Itarsi is about 200 km from Education Park. The cost of transportation from Itarsi to Education Park was quite high. Another thorough gentleman, Mr. Ambrish Kela provided us with two vehicles, which we used for transporting material from Itarsi.

Locals were identified who would assemble lamps, the training was conducted, the Solar lamp assembly started, and then we started going to schools to tell them about the Solar lamps.

From the beginning, the plan was to provide the lamps to the students for INR 200 (USD 3 approx.), whereas the total cost of the lamp was INR 400 (USD 6 approx.). The remaining INR 200 per lamp was to be raised by other means. My plan was to raise INR 200 per lamp not only from donations but also from well-to-do school students, with an idea of encouraging brotherhood.

During Diwali, a festival of lights in India, I appealed people to gift a light to fellow students in rural areas. For this appeal, I decided to go around and spread awareness, I addressed nearly 10,000 students in various schools and colleges.

Now, we started approaching schools in Khargone district and encouraged students to buy Solar lamp for INR 200, whereas the actual cost was INR 400. Though we were constantly visiting schools but somehow desired results were not coming in. Several visits to schools and discussion with teachers were not proving useful in achieving the purpose. Students were not buying the solar lamps at INR 200.

HOW THE TABLES TURNED?

Now, I had to think of some good strategies to promote it, as I was very clear in my mind that the lamp is very useful. I could see it impacting the lives of people. I thought that the lack of awareness about the lamp, it's power in terms of light and the cost of the lamp were bottlenecks. To address this, I thought of educating students' parents about the lamps. To reach out to the villagers, we started doing night campaigning, using small loudspeakers, announcing and explaining the benefits of Solar lamp.

When I attempted this for the first time, there was hesitation in me: *How to start? What to tell villagers?*

I gathered some courage and went to a small shop, switched on the lamp and started talking to shopkeeper. He heard me out, showed some interest and eventually wanted to buy the lamp. This gave me some confidence. Then, the real campaign started.

I would stand in a public place in the village, switch on 3-4 lamps and invite people for discussion. Usually, 10-15 people gather in no time. I would tell them about the lamp, that it is available through schools and that it is subsidised. I trained the team also for what to speak in public, what dialogues to use, and how to tell about the cost of the lamp. Normally, people are hesitant to talk in public, so was the case with other team members.

Also, in order to give villagers an experience of Solar lamp during the night hours, we gave some lamps to teachers for trial purpose. I have seen that many people would visit the house of teachers to see the lamp and amount of light it gives. I thought it was good way to promote the lamp.

While all this was happening, I started sleeping in the campus of Education Park, which itself is located in an isolated place, thus not having any proper arrangements to stay. I was really happy living in the school campus, all alone, as it gave me time to think and focus on promotion of Solar lamps. Following me, several other team members also decided to sleep in the campus. It became fun. We would wake up in the morning, discuss the plan, the team would go to schools in the day time for campaigning, and after dinner, we would review the day's progress.

After two weeks of hectic campaigning and awareness creation for the lamp amidst all stakeholders: students, school teachers, parents, villagers, village leaders, etc. there was not much traction shown as far as the sale of the lamps was concerned.

I questioned myself, *"Why no one is buying the lamp?"*

Is something wrong with the program design? I took a pause and felt that we have done everything from our side.

I told the team, if even now, we do not get any order for the lamp, there is something wrong with the lamp, or there is something wrong with the thinking that the Solar lamp is required in villages.

I told them, *"Let's wait for few more days, else we will wind up the program."*

FINALLY, SOME ORDERS CAME OUR WAY!

Somehow Sunday passed, on Monday morning at about 11 am, someone came running to me and said, *"Sir, I got a call from a school. A teacher called to say that he has collected money from 10 students and want us to deliver the lamps to the school."*

I reacted *Wow!*

What a moment it was to hear this, we all jumped in joy! Soon someone on his bike went to deliver the lamps. On the same day, we got two other calls requesting delivery of the lamps. This way we noted that the idea of the Solar lamp was not flawed, INR 200 was affordable and people paid even in this tribal district. Assemblers and distributors of the lamps, who came from villages could perform their task properly.

The ball started rolling and many calls started coming in, Solar lamps started going in the hands of students. The data regarding the students was collected every day. Some students from IITB helped in creating a webpage regarding our Solar lamp campaign. The most interesting part of the website was a counter on the website, that showed how many students have taken the lamp. The number will get refreshed every day. So, the counter was something very interesting and people started following it.

LIGHT AT THE END OF THE TUNNEL

Followed by the above developments, I got a chance to share the concept of the program with Principal Secretary, Energy, Government of Madhya Pradesh, Mr. S. R. Mohanty, who found it to be quite exciting. After this meeting, the Government of MP announced a contribution for 10,000 students. The officer would check the Solar lamp counter on the website and tell me that he is happy to see the progress of the program, live.

Within 4 months, more than 21,000 students bought the lamp for INR 200 in the district. It was a success story. I was thrilled with the numbers, so was everyone else involved in the program. Though I wanted to reach out to the entire district, having just one assembly centre was a limitation, considering the size of the district.

We saw the use of the lamp not only for study purposes but for many other purposes, in the kitchen, for milking cows, for going to the farm, etc. The Solar lamp became part of their lives.

This was just a beginning. This tiny lamp sowed the seed of something big to happen. The experiences that I have gained were enormously valuable. The project demonstrated that the localization of Solar energy works well, much better than any other way of promoting Solar energy in rural areas. The localization is much more efficient and effective, it can work at a great scale and with great speed.

When the whole thing was set up. I thought the job was done and I got back to IITB to take it forward.

This is how Chetan's journey in a big, bright world of Solar began and progressed.

This was my real experiments with Solar energy deployment, which showed me the SOLAR truth, the way of implementing successful Solar programs.

PART - D

WIDENING VISION – SCALING UP THE SOLUTION

I strongly feel that the problems related to basic necessities of life in the rural areas including energy access, clean cooking, clean drinking water, quality primary education, crop preservation, food processing and health access, all these problems still exist, not because of the lack of technological solutions, but because of the lack of involvement of local communities' in problem solving.

– Author

CHAPTER 10

LEARNING THROUGH MILLION SOUL

After coming back to IITB from the wonderful experiment of *'Localization of Solar Energy,'* I spent the remaining part of my sabbatical (year 2012) in writing my book, creating training content and preparing a project proposal that would have the potential to change the direction of Solar energy generation and consumption, in the world!

As a part of the agenda of my sabbatical, I started writing another book on Solar energy. My first two books were quite successful, which were meant primarily for engineers and researchers. Now, the idea was to write a book for technicians and practitioners, which could be used as a workbook, covering topics like: How to design a Solar system without knowing the physics behind it?

How to install a Solar system? How to maintain a Solar system? This book was written in a very simple manner like an 'Easy to use handbook,' with several small calculations and thumb rules. The book was titled as, "Solar Photovoltaic Technology and Systems: A manual for technicians, trainers and engineers".

When the book was published, it built a wide readership base in no time. The book also got translated in five regional languages under the aegis of the educational activities of NCPRE.

I was enjoying the efforts that I was making to spread the word about Solar energy everywhere, for everyone through my books, video lectures, invitation lectures and short-term training courses on various topics for participants from industry and academics. One of my ideas was to create enough reading and training material in the field of Solar energy so that there is no dearth of material to learn Solar technology.

My experiment with the localization of Solar energy and the Solar lamps were quite successful. It strengthened the idea of localization, broadened my vision and opened the doors for infinite possibilities of localization.

What more could I expect from such experiments?

RIGHT-TO-EDUCATION VIS-À-VIS RIGHT-TO-LIGHT

Around 2012, a large number of villages in India would have either erratic, low-quality or no electricity supply. Census data across the world revealed that the situation was no different in many countries of the world, especially in developing countries.

A thought occurred to me while thinking about such villages. It was related to the Government of India's "Right to Education" to children below 14 years.

I pondered, *"How in the absence of reliable and quality supply of electricity, especially during evening hours, students can exercise their right?"*

This scenario would certainly affect education, which was clear from the high dropout rate of students from rural schools.

Why shouldn't students have "Right-to-Light" as well?

Yes, Right-to-Light.

I thought our Solar lamp experiment performed with rural communities could be one effective way of providing Right-to-Light. This became the idea for the next step, providing Right to Light to every child, and, if I am free to dream, I dream it for every child, around the world.

When it comes to Solar energy, particularly providing solutions to rural areas, every officer around the world wanted to do a pilot project of his own. All government officers want to thrive on precedence, but do not want to take the risk. Unfortunately, there are many not so good officers, who take the risk for wrong purposes and succeed. But, on the other side, there are many good officers, who wouldn't take the risk and no path-breaking work would be done.

ONE MILLION LAMP PROJECT, BIG ENOUGH, NOT TO BE A PILOT

Somehow, I had developed a dislike for the word "Pilot Project". Solar Lamp program was running in the world for several decades, then why one should execute it like a pilot project? Based on my experiment with localization of Solar energy, I wanted to do a project on such a scale, after which no one would term it as a pilot project. This project should prove the point beyond doubt. I decided to provide Solar lamps to 1,000,000 or 1 million students, the number big enough not to be addressed as a pilot project.

Now, we had to decide the timeline of the project. I always believed that the rate of providing the solutions should be greater than or equal to the rate of growth of the problem. If the speed of providing a solution is slower than the speed of the problem growth, one would never be able to solve the problem. Its simple maths!

Many policymakers and executioners do not understand this to put it in practice. I thought of providing 1 million lamps in just one-year time, which was large in scale and speed both. In fact, some officers dissuaded me to plan a project of this stature. According to them, professors needed only small demonstrations to prove the point.

Based on my last experiments with a Solar lamp, the financial model that I proposed for the 1 million lamp program was accepted by many. I proposed that one-third of the cost should be borne by the government, one third from the society and one third by the students, thus equally sharing the responsibility.

This model was liked by Mr. Tarun Kapoor, an officer in the Ministry of New and Renewable Energy, Government of India and by social organizations like Tata Trust and Idea Cellular. It took some time from the writing of the proposal to actually securing funding for this very ambitious target of 1 million lamps in 1-year time, with the model of localization or through involving local communities. But I was lucky to get funding to implement this dream project. By early 2014, the dream came true, the project started, hence began my another journey in the field of Solar with another experiment.

SOUL FOR EVERYONE

I thought of giving a nice name to the project. Inherently, I have a penchant for creating good symbolic abbreviations. I found it simple here too, organised all keywords of the project on board and arrived at a meaningful name by making various combinations. In this way, the Solar Urja (Energy) Lamp or SoUL became the abbreviation of this new Solar lamp project.

Though this Solar lamp was primarily designed for the study purpose, from my earlier experience, I knew that the lamp is going to be used for several purposes. The name SoUL fitted well, as the lamp touched the life at the core, besides providing Right-to-Light. I liked it personally, so did everyone else. Later in the project, I created several abbreviations, which you will come to know later.

From the beginning of the project, Prof. N. C. Narayanan and Prof. Jayendran V. played an active role. We would have a series of discussions over a cup of tea on various aspects and planning of the project. Especially, Prof. Jayendran later became an integral part of every decision and action, thus played a key role. Both are good human beings, clean at heart, dedicated to working and fun-loving at the same time.

While designing the SoUL project, we ensured that the Solar lamp was for all students and not just for a selected category of students. We decided that every student, whether having electricity connection at home or no, studying in government or private school, rich or poor, everyone should have the opportunity to buy this Solar lamp.

Following this, we took block/taluka as a unit of intervention wherein all the students were approached. We called it saturation of the area. It resulted in a high density of beneficiary in the area and helped us in reducing operational cost.

As per the plan, about one-third of the cost of the lamp was to be borne by the students.

One important reason why everyone should have the opportunity to get the Solar lamp was, that we found in the rural areas even if there was an electricity connection, the amount of light on the floor (in villages normally students sit on the floor to study) that is required for reading was not sufficient. As per the international norms, the required light level for study purpose is 150 lux but our study of over 2000 households revealed that typical light level is only in the range of 15 to 20 lux in most rural households. Hence, I felt that this Solar lamp will be useful for everyone.

Some local politicians and government officers would insist on providing the lamp free of cost to a select category of students. I defended this everywhere with the argument that when students pay, they get the ownership, which results in better care and longer use of the lamp. I had experienced this from the previous project that if the need is there and the lamp that we are providing is good, everyone, even poor people are willing to pay.

In our research, we found that only 0.30% of people had difficulty in paying their share of the cost. About 99.7% people did not feel that the cost of the Solar lamp was high.

This was an important lesson, soon it became useful for learning in designing other Solar programs.

BEING LOCAL AS WELL AS GLOBAL

Since 1 million Solar lamps were going to be there in the field, therefore it was very important to ensure that repair and maintenance are provided for. We added another feature: not only the local communities will assemble and sell the lamps, but also repair the lamps in case of a fault. In this particular experiment, we provided 1-year free warranty for repair services to every student, therefore we had to establish repair centres everywhere.

These repair centres would only be viable if there were enough number of lamps sold in the area, as too many repair centres would have the significant overhead cost of operations and too fewer repair centres would not provide effective repair services. We devised a strategy that a cluster of about 3000 households should have one repair centre, for this a basic maths suggested that after one-year free warranty, there would be enough customers for the repair centre so that the centre sustains on its own. The creation of repair centres added to the confidence of students and their parents to buy these lamps.

Now, not only assembly and sale but also repair and maintenance of the Solar products was done locally. Soon we realized that the local repair centres have to be the base of the promotion and sustenance of any other Solar solutions in communities.

The next challenge was to find a way to the efficient execution of the project, 1 million lamps in one in a year's time, that too without adding significant operational cost. All the components of Solar lamps in the form of kit were centrally procured and provided to local communities.

This helped in reducing the cost of the lamp kit. In terms of other operations like Solar lamp assembly, its sale and repair, we formed clusters of villages in the sub-districts. Now the execution of all operations became efficient, right from identifying a local community, their training, the supply of raw material, assembly of lamps, and quality control to the repair of lamps.

We also devised a strategy to monitor the quality of the components and the Solar lamp at various points in supply, assembly and repairs, this was another important operational issue, as many people would doubt the quality of the lamp being made in rural areas. Here all the quality checks were done by the local assembly and repair persons, except for the factory and the laboratory.

In a way, the formation of clusters led to an economy of scale as it reduced the cost of operations at every level. This design of the project also provided us benefits of being local (localized operations, low overhead cost and faster service) and of being global (cost-effective, high-quality components).

OPPORTUNITIES & CHALLENGES IN EXECUTION

Beyond doubt, reaching out to 1 million students in one-year time was going to be a phenomenal task. Since IIT Bombay did not have any connection with the rural areas, we made a strategy to partner with other self-help groups and NGOs, which were working in that particular area and had a connection with the local community. This also turned out to be a good approach.

Through these organisations, we could easily connect with the people of that area, therefore efforts were not required from our side to reach out to communities and build a relationship and establish faith. By forging a partnership with local organizations, in no time we could establish a very wide network, spreading to nearly 10,000 villages. This network was an important aspect for completing the project in stipulated time.

Prof. Jayendran, with his specialization in Industrial Engineering and Operations research, utilised his experience here. With his help, we also designed a highly efficient tracking mechanism for material movement, monitoring number of beneficiary students, monitoring number of lamp repairs, which component got repaired, the financial inflows and outflows and various other aspects of the operation. He was simply fabulous in all these. With his inputs, not only that we completed the project on time, but also collected accurate data electronically for each and every one of 1 million students.

It was the operation of enormous scale accomplished with high accuracy, especially when it was managed by an academic institution (IITB). We, being an academic institution had an important vertical of research activities while executing the project. Through these activities including surveys for data collection and group discussions etc. regular inputs kept coming to us directly from the field, which resulted in further improvisations in lamp design and processes. The direct touch and interactions with locals showed us the path of improvisation as and when required.

Even with all this execution plan being in place, still reaching out to every corner, providing training to the local communities, ensuring the material reaching to remotest of the locations and working towards setting up of repair centres etc. was easy. Many times, during the rainy season our team had to walk through the knee-deep waters or take a boat to reach their locations.

I had been motivating the team that it is important to finish the project in one-year time so that we can demonstrate that with this decentralized model, one can achieve such a large scale and the high speed that is required to solve the real-life problems.

I strongly feel that the basic problems related to basic necessities of life in the rural areas including energy access, clean cooking, clean drinking water, quality primary education, crop preservation, food processing and health access, I believe strongly, that all these problems still exist, not because of the lack of technological solutions, but because of the lack of involvement of local communities' in problem solving.

I was lucky to have an efficient and dedicated project staff team with me. They could associate with the problem and viewed it as an opportunity to solve the problem. I am really indebted to each and every one who participated in the project and helped in implementing it successfully and more importantly created experiences that could potentially change the manner in which the world generates and consumes energy.

Except the time when no funding was available due to procedural issues, when work had to be stopped due to lack of material, holiday in schools or festival breaks, we could actually finish the project, i.e. providing 1 million Solar lamps to 1 million students, in just 11 months, actual days of operation of assembly and sale ran only for 11 months.

With this 1 million SoUL project, we have reached nearly 10000 villages in various states of India. While the project was running, I had the opportunity to visit several villages, several Solar lamp assembly and repair centres. I had an opportunity to interact with people both men and women involved in the project. Every time when I went to meet them, I was instilled with fresh energy. Every time they would interact with me, I felt empowered towards my resolve of doing this type of project even on a bigger scale.

When I heard their experiences, when they narrated the benefit of this simple Solar lamp project, I could see and feel the involvement of the people, the money that they were earning, the confidence that they were showing in working with the technology, handling of the technology, the empowerment of women, the growth in their stature in the society and the respect that they earned from their husband and family members, were real eye-openers. Several people reported that they won Panchayat elections due to the Solar lamp, as they would become famous in the region. There were many stories of how the use of Solar lamp changed their life.

LAMP TO THE RESCUE OF A LADY IN LABOUR

In one such case, the lamp proved helpful in driving a pregnant woman 18 km in the dark night for her delivery. This is an instance of the hilly tribal village in Barwani district of Madhya Pradesh, during mid-night, a pregnant woman was to be taken to the hospital. In the absence of an ambulance, the villagers could find an old four-wheeler, but the driver refused to drive in the dark night in a hilly region, as he did not have headlights in his vehicle.

Fortunately, few villagers quickly brought the SoUL lamp distributed in that village, tied the lamps in front of the four-wheeler and in the light of SoUL, they drove the pregnant women for delivery.

SoUL was also used by villagers in milking cows, going to the farm, cooking in the kitchen, going to the temple, and even at social gatherings. The SoUL was used in all activities from birth to death. In fact, I was pleasantly surprised to find that the number of deaths due to snake bite in their villages had gone down after they got SoUL. I have also seen that the Solar lamp was used in a few other disasters in Uttarakhand and Kerala due to its portability, simplicity, and low cost.

Can any other product have such a wide range of applications? It only tells us how powerful a small device could be. For me, this Solar lamp was symbolizing the transition in the lives of people.

How can governments not take up such an approach for providing technological solutions, not only in the field of energy but for all other fields where solutions to basic necessities of life have still not reached to all communities across the world?

How can governments not see the enormous potential which localization of technological solution holds?

Why can't we bring the technology in a simplified version and give it in the hands of local communities? I could see this by way of localisation, we could solve many problems of the community including that of livelihood, electricity supply, education, clean drinking water, women empowerment and many others.

While visiting these Solar lamp assembly centres, I would ask them: *Are you happy?*

I used to tell them why it is important to have a smile on the face, my training as Happiness Program teacher of Art of Living, used to be very handy in such close interactions with the community. I used to sing with them, laugh with them. They would, in return show a lot of affection, they would make special arrangements for me to be received at the centres, they would decorate the centre for my visit, sometimes they would shower flowers on me, this way all these interactions would make the bond between me and community stronger.

WHAT BEYOND SOLAR LAMPS?

During my meetings with the local community in the Assembly centres, they would always show their excitement of being part of the project, the ability to earn money, they would have all the praise for the project, but towards the end of the meetings, or sometimes in private, they would pose a question, an uncomfortable one, *what would happen to us when this project is over?*

Hence, I started thinking very seriously what could we do beyond Solar lamp project?

I used to joke about Solar repair centres, especially in my lectures in cities, that these days many souls are in trouble, and if you see any troubled soul, bring him to our SoUL repair centre. This way indirectly we were helping many souls to be in better shape.

Such is the power of SoUL!

Today, after several experiments with Solar energy, and several years later after successful implementation of the 1 million lamp project, I feel that this Solar lamp is required even for American and European students. Training on the use of Solar energy is as essential as studying itself, as only Solar energy can become the base of sustainable life on the planet.

CHAPTER 11

THE TRANSFORMATION FROM SOUL TO SOULS

The Solar lamp project became an important tool to generate livelihood, provide skills to the people, create awareness on Solar energy, giving the first-hand experience to people of the use of Solar energy, generating confidence in the Solar technology, therefore needed to continue and spread to all the regions. But, immediately after starting 1 million Solar lamp project, it became very clear that Solar lamp project could only be the beginning, it cannot be the end.

The whole idea of doing 1 million Solar lamp project in one year, involving local communities in all operations of the lamp assembly, sale and repair was to establish the fact, beyond doubt, that this localization model can be used for large-scale and high-speed operations.

While thinking of the next step, I had to take care of two things; Scale-up higher than 1 million and plan a project that goes beyond Solar lamps.

What could be the next large scale for SoUL project: 2 million, 5 million or 10 million? I thought of implementing the project for the whole country. Anyway, the idea was to provide "Right-to-Light." My argument was simple if covering and providing a Solar lamp to the entire population takes long, many students would graduate from school but worse, many millions would drop out of school, and we can never claim success for the project. Hence, I felt the need to cover the entire country in just two years' time. The scale I could comprehend, I could imagine this happening all over the country parallelly, I could see the localization model working very well. I could imagine everyone studying in bright clean light of Solar light. But for many, the scale was too daunting!

APPRECIATION & GRANT FROM THE MINISTER

We invited Mr. Piyush Goyal, the then Minister of New and Renewable Energy to announce the completion of 1 million SoUL project. He kindly accepted our invitation. At that time, he was talking about electrification of all villages in India, and here I was going to tell him about the Solar lamp program for all villages in India. Indeed contradictory!

Why would a Minister planning to provide electricity to all villages also provide partial funding for providing Solar lamps to all villages? That too a tiny Solar lamp with only 1-watt power of the lamp. But I was convinced that we need a large scale Solar lamp program, as it is not only about providing Solar lamps, but it is about local skills, local livelihood, local assembly, local repairs, confidence in technology, empowerment of women, education, etc.

When the Minister visited IIT Bombay, I showed the entire process of operation of 1 million SoUL project, demonstrated the assembly and repairs by women from Rajasthan.

When the Minister asked these women, *"How much time it takes to make a lamp?"*

One of them replied, *"It takes just 20 minutes, shall I show you?"*

And she started demonstrating, soldering with steady hands. Seeing all this, the Minister was amazed by her confidence and skill. I think this demonstration must have cleared all the doubts in the mind of the Minister about the power of localisation of Solar energy.

I made a presentation to him with a vision to provide 100 million Solar lamps across the country, especially in areas with a shortage of power. And then, in his speech, he praised me for the efforts, and when he announced that the government will support providing 100 million lamps and provide partial funding for the Solar lamp program of Prof. Solanki, all of us present in the hall were stunned, including our Director.

He also invited me to Delhi to formulate the program and launch it. Wow! The entire team was thrilled, none of us expected such a positive response and quick action. Soon, I was in Delhi along with Prof. S.B. Agnihotri to discuss the plan. Due to operational reasons and speeding of the project, the plan had to cut down to 7 million lamps. By any stretch of the imagination even reaching out to 7 million students would be a great step forward.

ONE MILLION TO SEVEN MILLION, NO MEAN FEAT

Soon, we sat down to implement seven times larger project in only double the time of the last project. Prof. Jayendran stood beside me in the entire planning and implementation of this new 7 million Solar lamp project. It was definitely going to be much more challenging, but at the same time, more exciting as it would open up several other possibilities.

In this project, we mainly worked with Self Help Groups (SHG) working under the National Rural Livelihood Mission (NRLM) of Govt. of India, which was meant for women in rural areas. In this way, all the local operations in the project Solar lamp assembly, sales, repairs, data entry were performed by women only. We started calling them 'Solar Didi', in English Didi would mean Solar sister. This model of working with SHG women was going to ensure the sustainability of Solar technology in a rural area as they have deep roots in the communities around. It also established the synergy between the efforts of NRLM and SoULS, resulting in livelihood generation of SHG women.

The implementation of 7 million SoUL project is more complicated than earlier 1 million SoUL project. There were certain things added by the officers in the project, which I felt did not make much sense. Since government funding was involved, we had to accept it, later this created problems in the execution of the project.

The officers who do not have a sense of the situation at the ground level, and do not understand how the project was being implemented on ground, they often tend to make wrong decisions. I have seen this happening, so many times while working closely with government officers, I could see the impact of a wrong decision in the field.

RELUCTANCE BY GOVERNMENT OFFICIALS IN THE EXECUTION

Though I always tried to push the project in the zeal to provide a quick solution to people, only a few officers would show the inclination to ensure the timely completion of the project. Even if they wanted to push, their own bureaucratic processes will stop them. A new agency was employed to do procurement of the lamp kit, which brought their own share of problems and worse of all, delays.

To me, it explained, why it takes so long time to bring the solutions to people. I guess this must be true in every other country.

In fact, I got a shock of my life, when an officer told me, *"Why are you hurrying, people have waited for 70 years, can't they wait for one more year?"*

As the project implementation progressed, I became more and more disillusioned about working with the government. It became clear that no matter how much the government would provide the funding, it will be difficult to push the solution at a scale and speed that you want.

DECENTRALISATION VERSUS CENTRALISATION

Clearly, there is a different approach that many governments across the world adopt, they would like to centralize the solution, while the SoUL tries to decentralize the solution. Therefore, there was a lukewarm response from the government. As a result, I would start thinking of alternatives for bringing solutions to people through localization model. Even for this experience, I am thankful to the officers who forced me to think in new ways to find a solution.

As time passed and SoUL lamp reached more and more people, it started getting recognition. Because of its portability, the lightweight, long performance of 12 to 15 hours running on a single charge, easy handling, the SoUL lamp was used not only for study purposes but for many other purposes in villages.

The answer to the question from the local communities *"We do not want to sit idle, what we will do after the project?"* remained.

I was thinking if they can make a Solar lamp, why can't they make all other Solar products, at least simple Solar products? I thought with little more training they could definitely do this.

THE IDEA OF SOLAR SHOPS

One fine day, the idea came to my mind that the local communities can start Solar shops for selling all types of Solar products required in rural areas. It was like a seed germinating and becoming a plant, a seed had already been sown through SoUL. The trust-building among Solar Didis and the local community was already present due to our efforts. In this context, I thought that Solar shops will have good acceptance.

The plan was to provide them with more advanced training and knowledge of these products so that they can sell and repair also, a key aspect of localization. Thus, the opening of Solar shops or I abbreviated it to Solar Marts or S-Mart of "Smart Shops" added yet another dimension to our Solar lamp program. These Smart shops were run by S-Mart Solar Didis.

NOW SOUL BECOMES SOULS

Since the project was not only about SoUL or Solar Urja (Energy) Lamps anymore, I started thinking about the new abbreviation that can rightly describe the new intention. I went to the board, assembled all the keywords, and I could not believe that another abbreviation emerged, similar to SoUL, but it now described the approach appropriately, the abbreviation was SoULS or Solar Urja (Energy) through Localization for Sustainability.

It meant that we now look for complete Solar energy solutions and not only Solar lamp, it meant that localization is a key aspect of all Solar energy solutions. And not only do we look for the sustainability of the Solar solution in the narrow term, but the sustainability of life in a broader term.

One of the important learnings from my Solar experiments was the difficulty due to non-standardization of Solar products. There are literally hundreds of different types of Solar lamps available in the market, each one of them has their own design of the body, circuit, panel, etc. Due to this, the components from the different suppliers are not exchangeable. Hence, if you buy a product from one supplier, and if there is a fault, you are dependent on the replacement of the component by the same supplier. In our 7 million SoUL project, we have designed the entire lamp and asked every supplier to use the same design. As a result, any component can be bought from any supplier.

OPEN SOURCE HARDWARE

The lamp design, with several refinements, became quite robust. We have decided not to file any product design patent rather release the entire design on the public domain. This has marked the beginning of our efforts to create an open-source hardware design for all other possible Solar products. My thought behind this was the same, which I had, while creating the teaching and training material, that no one should be limited by the knowledge of Solar technology and Solar products.

Anybody who wishes to promote it and run their livelihood by selling Solar products, they should not be limited by the knowledge of the technology. Hence, I also started focusing on the creation of open-source hardware, for at least all commonly used Solar products.

When some Solar shops got established, there was an issue of procurement of the raw material for other Solar products and their supply. Since the Solar shops are small and their requirements are small, no vendor is generally willing to supply the small quantities, that too in remote rural areas. So, what could be the solution?

Also, I thought, if these women, Smart shopkeepers keep buying material from the supplier from the big town, the money from the local area would keep flowing out to bigger towns. As the material contributes to the major cost of the product, significant money will go out of the local economy. The villagers earning would not be sufficient from only assembly and repair services.

"Can we not manufacture Solar products locally?" the question came to my mind.

I started thinking about the possibility of manufacturing Solar products in rural areas by locals. The simplest product that I could think of manufacturing in rural areas was a Solar PV module or Solar panel, probably due to my familiarity with it. I have worked on module manufacturing during my Ph.D. and within India, I have seen almost all the factories of Solar module manufacturing, so I knew the process in and out. With this thought, I set out for yet another experiment in the field of localized Solar solutions.

LOCALIZED MANUFACTURING

Though it was about the Solar lamp, I proposed this idea of the possibility of setting up a small Solar module manufacturing facility in one of the meetings with the tribal women in Dungarpur district of Rajasthan. The then Collector, Mr. Inderjeet Singh was also present in the meeting. The idea was exciting, but no one, including me, had an idea what it would entail. There was no way a big module manufacturing plant could be established in a rural area.

I was not aware of small machines, suitable to produce small size Solar module to fulfil local needs. Only after coming from this meeting, I started searching around and finding the machines and their cost. Soon, I put a business model to set up a Solar module manufacturing unit in Dungarpur. What happened there, was simply amazing!

PM'S AWARD FOR INNOVATION

The Dungarpur Solar lamp project was funded by CSR funds of Idea Cellular which provided us with the flexibility to experiment. I asked the women to sell the lamp at a higher cost, and as a result, students have to contribute about 40% of the cost, but they got the lamp still with a subsidy. The additional cost that women would collect from the higher sale price, would be utilized as a corpus for setting up PV module manufacturing plant.

The women in the region could sell nearly 40,000 lamps, but with great difficulty. Everywhere else, the lamp was being sold at a lower cost or higher subsidy. Selling of 40,000 lamps at a higher cost was an amazing achievement. This created a good corpus, but not sufficient to set up a manufacturing plant.

While all this was happening, the collector Mr. Inderjeet applied for Prime Minister Award for this project. I witnessed the award selection process, as it went through six steps, which I was part of and answered questions at various steps. The last step was interaction with Cabinet Secretary of India, which was scheduled at 10 am through video.

When the result was announced, the Dungarpur SoUL project was declared the winner of the Prime Minister's Award for Innovation, the award was received by the Collector and given by the Prime Minister Shri Narendra Modi. Due to this award, almost every Collector in the country came to know about the SoUL project.

SETTING UP OF DURGA ENERGY

In the process of setting up this manufacturing plant, many things happened and many people helped. An old school got allotted for the factory, the local government repaired the building and made it ready, some more CSR funds were provided by Idea Cellular and Government also pitched in some funds. Soon the company was registered, first of its kind, probably in the world where entire ownership was with women only and women were going to run the show. Many bureaucrats doubted that such an initiative can be successful. Many would also doubt the quality of the Solar panels produced from this factory and so on, but I was confident.

After finalizing a vendor, we went to China to check the machine, before its dispatch, and when machines arrived, training of all women happened, inauguration happened and soon the manufacturing plant was up and running. Apt with the work, the name of the company was kept Durga Energy. This factory is now running for more than 20 months, of course with its own share of ups and downs.

It is difficult to declare it as a success, as it is yet to see its complete operation without any help from people outside. The local community is yet to sell all its products within the local area. Right support from the government and policymakers is required. Like, for local district needs of the government, priority should be given to the produce from the same district.

This simple approach can make an enormous difference in the efforts of every country to generate livelihood and employment opportunities for every citizen. This approach can help in stopping migration from rural areas to urban areas and all associated problems with it.

Good news is that such localized production and consumption of Solar products makes sense to many people, as I am writing this there are three other factories of this nature are being established in various parts. I have no doubt, this can be done and implemented in any other part of the world as well. The localized manufacturing can be the backbone of entire localization, it can be a magnet around which all activities of supply, assembly and repair of Solar products can take place. This can be a new way of providing quick Solar energy solutions.

Now, I thought how about touching lives in every household? Kitchen in the home is where all the action is. I wanted to rope in Solar energy in the cooking process which is one of the most important needs of life.

COOKING WITH SOLAR ENERGY

Where ever we hear about Solar based cooking; it is always a Solar thermal way of cooking. A typical example of this was the box type Solar cooker, you might know of. There is another advanced cooking solution called dish type of cooker, which has been there for several decades but could never become the choice of people, mainly due to the difficulty one goes through while using. The user has to follow the sunlight which makes precise control of cooking difficult. Several alternatives have been worked out, but the wide-scale proliferation of these solutions never happened.

I started thinking, why can't we make a cooking device with the function of the box-type cooker which is mainly used for cooking rice and lentil or any other food that can be cooked by simply boiling. It requires heat to be generated. Why can't we use Solar PV panel, which is very easy to use. It has also become cheaper. Why can't we convert this Solar electricity into heat and use this heat for cooking? Once we generate electricity, we can cook in comfort of being inside the house. I set a challenge that the cost of this Solar PV based cooking device should be equal or less than the box type cooker.

Scientifically sunlight for electricity generation and then using electricity to generate heat is not a good option. It is definitely less efficient than directly generating heat. But electricity provides many benefits, like the convenience of cooking indoors, control of cooking by push of a button, provides energy storage option that can be utilized in non-sunshine hours, one does not need to keep changing the position of the panel and therefore simple operation.

To me this made enormous sense, even at the cost of lower efficiency, as long as we get the same cost solution, efficiency anyway does not matter.

I guided the technical team to work on a solution which was simple, converting electricity into heat. Within a month of trials, we were able to cook rice and lentil. One of the key things in our experiment was to ensure a reduced loss of heat from the vessel so that almost all the energy generated by the Solar panel is utilized for cooking only.

We were exploring another possibility of generating heat using Solar electricity by using an induction coil. The key idea was if this can be done without storing electricity in a battery, then we can do day time cooking at a very low cost. The initial experiment proved successful. I was personally really excited about this.

OUR COOKING SOLUTION WINS A COMPETITION

Around this time, someone brought to my attention a competition organized by ONGC (leading Oil PSU) for designing a Solar cooking solution, as quick as LPG based cooking. I jumped at the opportunity, started developing Solar PV panel-based cooking solutions. The competition had tight deadlines and several stages. Soon we were fully engaged in the development of this cooking solution. The Solar technology lab got converted into a kitchen with potatoes, tomatoes, onions, etc. all over.

This was due to the reason that I set the challenge that if we eat lunch, it should be cooked on our own solution, else we will not eat. Hence, we had to put all efforts to ensure that cooking happens through our solutions. This really helped us to perfect our solution Within a few months as we were bound by the deadline of the competition. The efforts paid and we had a good solution in hand. We just played with the best technical knowledge available and using our experience we put together a solution.

When the result of the competition was announced, we won the first prize. Besides, we also got the opportunity to implement our cooking solution in a village in Madhya Pradesh. With the help of CSR funding from ONGC, we installed this cooking solution for the entire village called Bancha, in district Betul of Madhya Pradesh having 74 households.

How could I miss the chance to localize the solutions? We trained the locals and complete installation of the cookstove was done by local students.

We provided them with training and made them part of the entire installation process. This enabled us to get the work done quickly in a remote area. It also created the opportunity to have someone local for any minor repairs. Since then we have been monitoring installations of the cookstoves, except for a few issues, all cookstoves are working fine and people are happy cooking on our Solar solution.

SOLAR CHAI THELA

Once you have a Solar cooking solution, the sky is the limit for its application. I thought of Solar chai or Solar tea stall. The same cookstove can be fitted on a cart, with Solar panels installed on the top and battery and other electronics items can be put below the base of the cart. This way we can make Solar chai cart/*thela*. We successfully installed a Solar chai *thela* in IITB campus, where everyone enjoys a hot cup of tea. I can see that this concept can easily be replicated for creation of livelihood opportunities for thousands of people in many cities.

This Solar cooking solution opened many doors in our efforts to provide Solar solutions to the rural community. This success gave me the vision to think of complete Solar solutions for communities, all of which can be provided with Solar. All the energy needs of the local community through Solar looked feasible and I had the first-hand experience of developing and providing all these solutions.

My vision of Solar solutions, in general, became wider, the entire canvas of possibility of Solar solutions, their capacity to change the lives of people, the problem that can usually occur in Solar solutions, good and bad experiences that I gathered, what can make speedy operations and what slows down, when and how the funding can become a limitation, and so on.

All my experiments with Solar are fresh in my mind, I can draw them on the canvas. This breadth and depth of experiences are as valuable as a treasure. My Solar experiments have taught me many things. These SOLAR truths have shown me their potential and limitations. As a way forward, Now, there is a need to bring localized Solar solutions with their potential while overcoming the limitations.

Many times, I feel, how lucky Chetan is to have the opportunity in life to go through all these experiences?

One can only be grateful to the visible and non-visible energies around us.

CHAPTER 12

THE SOLAR ECOSYSTEM "BY THE LOCALS, FOR THE LOCALS"

After experimenting so much on providing Solar solutions to the communities, I would ask myself, *"What would be the complete picture of Solar solutions?" Can I draw it on canvas?*

Only a little thinking would tell me that only Solar lamps or Solar shops or even a Solar module factory cannot be the solution. There are other elements of Solar solutions that have to be in the right place like the absence of funding can impede the widespread use of the Solar solutions, without appropriate Solar technology, the cost of the Solar solutions may not be affordable. Besides the lack of awareness among the community can make it a non-starter whereas absence of appropriate policy formation by the government may be detrimental to the promotion of Solar solution. This way the unreliable Solar solution can demotivate any user, without the view of sustainability, one may not be motivated to use Solar solutions.

Technology, skills, awareness, policy, funding, reliable products, availability of solution provider, manufacturing and users, etc. are the important elements that have to be part of the canvas of the Solar solutions. After all my experiments with Solar solutions, today, I view them differently. Among all these elements, I believe, if one tries to provide or work on only one or few elements of the Solar solutions, it may work in the short-term but it would be highly unlikely that the solution would work in long-term and sustain. The history of providing Solar solutions in a decentralized manner, in India and around the world, tells the very same story.

LOCALISATION HOLDS THE KEY

Starting with small Solar lamp experiment way back in 2012, implementing large scale Solar lamp program, starting Solar shops, Solar factories and developing several other Solar solutions, now, I have come to understand that the entire Solar ecosystem is required to ensure that Solar energy becomes part of the sustainable energy solution. Not only this, but I am also very clear through my Solar experiments that this Solar ecosystem has to be managed by local communities, for local communities.

I used my skills to create another abbreviation, Solar Ecosystem by the Locals, for the Locals or SELL. I found this abbreviation also very appropriate, as it is not possible for any government in the world to provide Solar subsidy to each and every household. There has to be a model, where Solar solutions get promoted through a market mechanism.

The governments can only afford to provide support to a small portion of the population, just to kick-start any solution, provide initial push and once the solutions get momentum, the government can move out of it. I believe that the government has done its part, for most of the population today, Solar solutions are affordable and the government can move out of it.

In my experience of today, particularly in India, I see that most of the government schemes and its intervention, in the form of policy and subsidy, is proving detrimental in promotion and growth of Solar solution.

The Solar Ecosystem by the locals, for the locals model, on the other hand, can bring many benefits to both people as well as governments around the world. Under the SELL, there would be several small-scale manufacturing units established in a given geography. These units, as much as possible, would produce the components of Solar solutions in local areas. These components would then be used by local Solar service providers and shop keepers.

The service providers and shop keepers would provide all types of Solar solutions together with their repair and maintenance services, to the households and other users in the local geography. I envisage that the SELL would also include a local training center as well as the financing facility. The training center, in collaboration with local academic institution would train the required manpower and would provide upgradation in the technical solutions.

The financing facility will provide loans to those who would need such support for installation of Solar solutions, including Solar shops and Solar related factories. In this way an effective and efficient localized Solar ecosystem can provide sustainable Solar energy solutions.

Governments around the world can play a key role in setting up such SELLs and by creating appropriate policies that would promote such a localized model. I believe that if the SELL model is adopted, subsidies and grants are not required anymore for the promotion of Solar energy solutions. Governments around the world need to enable setting up of SELL, without providing with any subsidy as grants for promotion of Solar energy.

PART - E

SUSTAINABILITY OF LIFE – ONLY WITH ENERGY SUSTAINABILITY

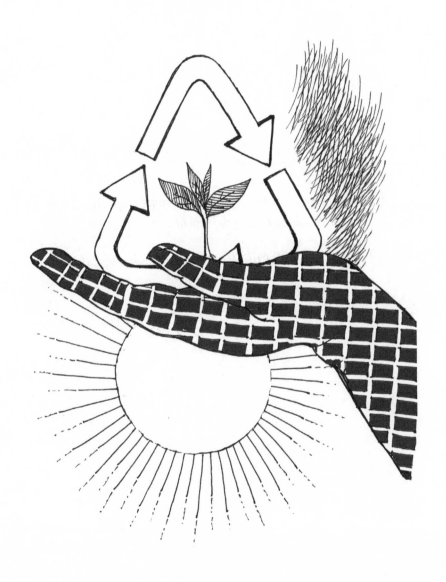

The Stone Age didn't end for lack of stone, and the oil age will end long before the world runs out of oil.

– Saudi Oil Minister Sheikh Ahmed Zaki Yamani

CHAPTER 13

WHEN I STARTED UNDERSTANDING THE SYMBOL '=' FOR SUSTAINABILITY

My experiments with Solar energy have not only taught me the ways in which Solar energy solutions need to be implemented but have also widened my vision. These experiences have led me to understand energy sustainability in the wider context of the overall sustainability of life on the planet. I understood how important the energy sustainability is, especially in the current context of climate change.

UNDERSTANDING SUSTAINABILITY

I have been reading about the word 'sustainability' from a very long time, and I understood the meaning of the word in a literal sense. It is not, until very recently when I started to understand the real meaning and the sense, the word 'sustainability' conveys. I started to understand what a terrible mistake it could be, not to understand the real meaning of the word it conveys, especially in a long-term context. I started understanding that living a human life without inculcating the real sense of sustainability is like crossing a busy road blindfolded.

I fear that more than 90% of people in the world, like my earlier times, only understand the literal meaning of 'sustainability'. Therefore, they are acting and living a life with this literal understanding, and putting human life on the planet in great danger.

> *"It (spreading out into space) may be the only thing that saves us from ourselves. I am convinced that humans need to leave the Earth."*
>
> – Stephen Hawking, Scientist

Humans have to leave the Earth, he explained, due to an array of threats including asteroid strikes, resource depletion, overpopulation, deforestation, the decimation of animal species, and the effects of human-made climate change, particularly rising temperatures and melting ice caps.

Are we heading towards *disaster*? Do we have to think about leaving the planet Earth? Let's discuss.

Several years back I was attending a marriage function in the family. Invariably, there was a lot of work to do and many functions to attend in a short span of time. So, I ended up working more and sleeping fewer hours. By the time functions got over, I was completely exhausted.

The question came to my mind *"Can I exhaust myself every single day like this?*

No Way!

If I try to do this, the body will break-down sooner than later.

"This is not sustainable," I thought.

The thought suddenly filled me with energy, I felt as if I was enlightened! It is quite simple. Understanding sustainability is quite simple. But, many times, simple things are hard to understand.

SUSTAINABILITY OF HUMAN LIFE ON EARTH

"=" or equal to is such an important symbol in the idea of sustainability, that if it does not get satisfied, it leads to several severe problems. In order to be sustainable, the left-hand side of the "=" symbol should be the same as the right-hand side. Say if your monthly income is less than the monthly expenses, then you are in trouble, particularly if it happens over a long period of time, then it would be fair to say that income-wise you are not sustainable. Your sustainability as a family unit would be severely impacted if you don't curtail your expenses.

Sustainability

Left hand side $=$ Right hand side

For sustainability, not only the left-hand side should be equal to the right-hand side, but this "=" relationship should hold true for as long as the life of the object on which we are applying this equation. If it is the question of our income and expenses, then it should be true for our lifetime. And, if it is applied to the existence of humans on the planet, then it should be true for millions and millions of years, isn't it?

The same "=" symbol can be used to compare many other things that we use and are an integral part of our lives. If it is a question of the amount of oxygen we breathe, then consumption of oxygen should be equal to the production of Oxygen and it should hold true as long as the life of the Earth, that is millions of years. Similarly, the amount of water we consume, the amount of soil we use, the amount of forest we use and many other essentials of life must have balanced consumption and production, not only for the time being and but for millions of years to come, if we want to live sustainably on the Earth.

Are we really living our lives in a manner that is required for the sustainability of life on the planet?

World Commission on Environment and Development, in 1987, defined sustainability as *"meeting the needs of the present without compromising the ability of future generations to meet their own needs."*

This is a very simple definition of sustainability and can be viewed using the "=" symbol. This definition tells us that our ability to meet our current needs should be equal to the ability of future generations to meet their needs. Is this true in today's scenario of the world? Does this equation hold true? Are we living in balance?

What do you (reader) think about it?

The answer is a big NO!

But arguably we are living in happier times where everything is available to us easily, with a push of a button, whether we want to eat, read, talk, sing, dance, walk, jump or even fly. Technology has made it all possible. We consistently go on replenishing things of our daily life including food, furniture, gadgets, clothes and even vehicles at a rate faster than ever before. And, most of our time, energy, money and other resources are spent on procuring and consuming these things. Today everything is centred around human aspirations which are growing by leaps and bounds.

Indeed, we are in luckier times when the standard of living is at its peak. Besides, basic food (*roti*), Cloth (*kapda*), House (*Makaan*), two-wheelers, four-wheelers, cars, well-equipped kitchens, TVs, Smart Phones, Computers, Laptops, Coolers, Heaters, ACs and other luxuries are in our easy reach. If a family can do with one car, it has one car for each member. If a family can do with one TV, it has one TV in each bedroom. Great!

But, hold on.

Have we checked if the symbol "=" is still holding true in this modern and comfortable lifestyle? Have we checked if our ability to meet these needs has an impact on the future generation's ability to meet their needs? Have we checked, if we are still in balance?

The answer is a big NO.

Is there any problem in the balance? The answer is a big YES.

We are in deep trouble!

EARTH OVERSHOOT DAY

The food, water, air etc. are continuously renewed on the Earth and living species are continuously consuming these resources. Let's look at the balance of the Earth as an ecosystem, what is its bio-capacity to produce resources or rather generate (or replenish) resources in a year and how many days humans take to consume these resources. If the natural resources that get generated in one year, get consumed in one year, then we are in perfect balance.

However, as we know, this balance has been disturbed by humans for a long time. In a year, we humans are consuming natural resources in quantities much more than they are getting generated in the same year. Hence, the symbol "=" is not holding true anymore, and, unfortunately, this is happening for a long time on our planet. We are already living an unsustainable life for long.

Each of Earth's resources is available in *finite* quantities, and they get regenerated each year in *finite* quantities only. If we go on using these consistently without giving a thought, what havoc it will create on planet Earth? Sustainability would take a big blow. I am not just saying this, there is a kind of book-keeping way of representing the disparity between human consumption and regenerated natural resources at a global scale, known as Earth Overshoot Day.*

Earth Overshoot Day is an indicator that measures the sustainability of life on Earth. Because the balance of the Earth has been disturbed, as we are consuming more natural resources than they are getting renewed.

Earth Overshoot Day is estimated by dividing the Earth's annual bio-capacity to renew natural resources by the world's ecological footprint of all species. The bio-capacity of the Earth is the capacity to generate biological material for humans and other species to meet the demand and the ecological footprint is the term which is used for people's consumption of Earth's resources for a given year.

Ideally, if we are in balance, the Earth Overshoot Day should come on December 31 of each year, meaning, whatever resources the Earth can generate in one year, we consume in one year. However, in the year 2019, the Earth Overshoot day came on 29th July itself. In other words, the resources that are generated in an entire year are consumed in just 209 days (days from 1st January to 29th July). The rest of the 156 days (30th July to 31st December) of the year, we over-use the resources from the Earth, we exploit nature, beyond its capacity...

...Hope the scenario is clear now.

There is a person who is running a business by investing some capital in it and earns an income. The person is free to use his income to meet his expenses but if he starts using the capital itself, the business will not be sustainable. Isn't it?

Similarly, we can only use the resources from the planet Earth which is regenerated (the income) but we cannot use the reserves of the Earth (the capital).

The water that flows in rivers due to rains, is the income of the Earth, we can use it as much as we want. But the water which is underground, and not getting recharged, is the capital of the Earth, we cannot use it if we are not recharging it. But, we are digging water and taking it up for use from deeper and deeper bore wells. The water table is going down in many parts of the world. It is unsustainable.

Take a pause, and think about the reduction in forest cover, increase in air pollution, degradation in soil quality, mining of fossil fuels or other resources that we use from the Earth. Think in terms of capital (reserves) and income (regenerated). Look what are we doing? Look how much reserves are we using? Look how unsustainable the life on the planet has become?

The reserves of the Earth (it's capital) do not belong to us only, but to all human beings and other species, even to those who are going to come in future to inhabit it. If we are using the resources from the Earth's reserves, it would not be wrong to say that we are stealing these resources from our own children and grandchildren. Stealing? Isn't it bad behaviour? Isn't it a shameful act of modern human beings, stealing right in front of their own children?

As rightly expressed by Greta Thunberg, climate activists from Sweden, *"You say you love your children above all else, and yet you are stealing their future in front of their very eyes."*

LIMITED CARRYING CAPACITY OF EARTH'S ECOSYSTEM

A system is a collection of parts working together to perform a particular function. The Earth's ecosystem has various elements like water, air, land, sea, weather patterns, forests which perform their functions to support the lives of millions of species. Similarly, the human body is an ecosystem of various organs and body parts, and they all work together to perform the tasks of the human body. From our experience we know that each system has limited capacity to perform a task, hence we should remain in those limits only.

If a bus is designed to carry 50 passengers, it will easily carry 50 passengers. But if it packs beyond this capacity, then it would result in stress on the system, leading to its breakdown in the long run, isn't it?

Similarly, the Earth is a biological ecosystem, it regenerates natural resources like air, water, food, land, etc. in some quantities, which are fixed. This way it has limited capacity to support the lives of people. There is no way, the Earth can support life of the infinite number of people for the infinite duration of time. If there are more users than the carrying capacity of the Earth, the sustainability of human life cannot be guaranteed.

How many people then the Earth can support? Or in other words what is the "carrying capacity" of the Earth? It is an obvious question that comes to our mind, but there is no easy answer to this question.

The number of people in the world is increasing, but the surface area of the Earth is not. The Earth is not expanding, even its resources that get regenerated remain the same.

Not only the number of people living on the Earth but also the amount of resources, each person is consuming, on an average basis, has become a bigger concern. Our consumption per person has been increasing enormously, therefore, it is resulting in a reduction in Earth's carrying capacity.

In this regard, we need to recollect Mahatma Gandhi's words, *"The Earth provides enough to satisfy every man's needs, but not for every man's greed."* I think, this statement clearly tells us that we cannot continue with our current (modern) lifestyle anymore. Gandhiji could view this, nearly 100 years ago. Today, things are much worse.

The relationship between the population, resource consumption per person and available resources on the planet can be given by the following equation:

$$\text{Population} \times \frac{\text{Resource Consumption}}{\text{Person}} = \frac{\text{Earth's Resources}}{\text{(fixed)}}$$

As per the above equation, if the population increases, the resource consumption per person must decrease as the Earth's resources are fixed. But, if the resource consumption per person increases, then the world's population must decrease. This is assuming that the total available resources required for the existence of life on the planet; the air, the water, the forest, the soil, the surface area, are fixed. And, I guess, no one can challenge this assumption.

Both, the overall population and consumption per person has been increasing, which are a big concern for the sustainability of human life on Earth. The population in early 1800 was only

1 billion, which increased to 2 billion by 1920s, and now it has become 7.3 billion. It is expected that by 2050 the world's population will rise to about 9 billion people.** How the planet Earth will support such a huge population?

It seems many people who are consuming a lot of resources on a daily basis have lost their mind, they have become insensitive to what is happening in the world. The only thing these people worry about is how I can accumulate more wealth, more resources?

Isn't this madness?

Yes, it is!

It is driving us towards disaster, sooner in future than later. I am personally, sometimes part of this madness, when I travel and stay in good hotels, I can see the kind of resources one person uses on a daily basis. When I recently travelled to the US and Europe, I could see how much resources were being used by people on a daily basis, it was, in a way, a very painful experience.

AREN'T WE BEHAVING LIKE INTELLIGENT FOOLS?

There is a mad race in the world for producing and consuming more and more. The world is busy setting up bigger factories, making bigger roads, making bigger machines, producing more cars and more machines. This is an ever-increasing trend, and by definition of sustainability, there is absolutely no way such an ever-increasing trend can be satisfied by the limited capacity of the Earth's ecosystem.

What are we doing on this planet?

"Any intelligent fool can make things bigger and more complex... It takes a touch of genius - and a lot of courage to move in the opposite direction," said E. F. Schumacher, a German Statistician.

Aren't we making bigger cars? Aren't we making bigger houses? Aren't we making bigger factories? Aren't we making bigger malls and bigger roads? It looks like that in our efforts to accumulate more wealth and derive more comforts, we are behaving like intelligent fools. Isn't it?

POSSIBILITY OR IMPOSSIBILITY OF SUSTAINABILITY?

I hope that after going through the chapter, the readers are in a position to write their own answer to this question. If I am able to convey correctly and if my arguments are logical, the most likely answer that you would have written down is that *"It is impossible to sustain this modern lifestyle on this Earth."*

If we look from the economic perspective, humanity is entering into a huge ecological deficit by consuming more resources than the capacity of the Earth to replenish each passing year.

It's not that it is an irrevocable situation, even a few percentage points changes can shift the date of Earth Overshoot Day by a good number of days. Are we ready to work towards it?

The population has multiplied manifold which is far from the Earth's carrying capacity. I could see this in all 30 countries which I travelled recently. I kept thinking about how the little empty area that is left on the planet would satisfy the needs of the population, now occupying almost all the spaces everywhere in the world?

We have to find some alternative solution to the rising population and urbanization soon. Climate change, scarcity of resources, an ever-growing population, and many more problems are knocking our door and now we need to open our doors for new ways of living where our future generations can enjoy the Earth's resources.

All the social and economic growth of humans of the last 150 years has been driven by the use of energy, mainly fossil fuels - coal, oil and gas. We need to find an alternative to these energy sources, as they were a major driver of the growth on one hand, but on the other hand, these fuels were also a major culprit of climate change.

For providing the same planet to our children and grandchildren that we inherited, for ensuring the sustainability of life on the planet, to ensure that humanity satisfies the "=" equation, we need to change. But then every change is challenging.

Are we ready to change? If not, then what?

Ref : *www.overshootday.org
    **www.Ourworldindata.org

CHAPTER 14

WE ALL USE FOSSIL FUELS – YET DO NOT WANT COAL POWER PLANT IN OUR BACKYARD

As you are aware that human evolution has been a very long drawn process. It has probably taken millions of years for humans to evolve, in shape and size that we are in today. If viewed zoologically, humans, also known as Homo Sapiens in primitive times, were culture-bearing, straight-walking and talking species who inhabited the Earth. The first evolution was probably noted in Africa some 315,000 years ago.

We are the only living members of what many scientists of the world refer to as the human tribe – Hominini.

There is ample evidence in the form of fossils to prove that humans were preceded for millions of years by other Hominins and other species Homos. With this started development of some kind of society, where everyone was hunter / nomadic who hunted for food the entire day and consumed it raw. Life was progressing very slowly in those times...

INDUSTRIAL REVOLUTION & FOSSIL FUELS

Hominis discovered fire while living in caves when they noticed that when two stones are rubbed with each other, fire is created. The curiosity grew and the fire was discovered, its usage evolved. This early discovery of fire provided several benefits to the early homininis, as it protected them from the rain, kept them warm during cold weather and they could cook hunted food on it, which made food tasty, healthy and easy to consume. Fire emerged as the most important component for the sustenance of human life on planet Earth and the following development.

But the major development which advanced the human societies by leaps and bounds was the rise of industry and manufacturing. Though the Industrial Revolution occurred thousands of years after the development of agriculture, beginning in the 18th century and consolidating in the 19th century, with the advent of coal, which started becoming the predominant source of energy.

This was the time when energy became a mainstream and most sought-after thing as human life started revolving around it. It changed the scenario completely as it led to one thing after another…

…this is how energy gained centre stage in the process of evolution of humans!

RISE AND RISE OF FOSSIL FUELS

In the modern world, everything is driven by energy. We have evolved our lives around energy generation and consumption as one of the key activities. Energy has become such an integral part of our lives that key parameters of social and economic development have a very strong relationship with the amount of energy consumption. Higher energy consumption on a per capita basis in a country would result in a higher increase in social and development indicators.

This is true for literacy rates, reducing hunger and poverty, improving health, reducing inequality, women empowerment, income generation, livelihood creation, industrialization, access to clean drinking water and so on.

Over the last 150 years, the increased need for energy has been supplied by increased exploration and production of fossil fuels. The production and consumption of fossil fuels have been ever increasing since the beginning of industrialization. Coal was one of the main fuels at the dawn of industrialization, which made way for oil, current-day petroleum, followed by gas, current-day CNG, PNG etc.

These fuels were and are being extensively used for carrying out our day-to-day activities like cooking, heating, cooling, cleaning, transportation and other energy needs. Termed as fossil fuels, these gained popularity in no time due to several merits like their condensed nature, ease of storage, transportation and cost-effectiveness.

Power plants started generating electricity which eased and advanced the scenario further as it was very easy to handle, easy to transmit thousands of kilometres and did not require storage. So, this composed the total energy scenario of the world. It became very important that wars were fought because of that. Not only this, economy, diplomacy, sociology, GDP and Development all based their manoeuvrings on energy usage patterns.

To cut a long story short, fossil fuels started ruling the roost!

THE WORLD RUNS ON FOSSIL FUELS

The amount of energy consumed by a nation is given in terms of Total Primary Energy Supply (TPES). As the term suggests it includes total energy consumed in the country for all applications including transportation, electricity generation, cooking and industry usage, etc. The TPES includes the energy resources that are produced in the country as well as which are imported from other countries.

The word 'Primary' in the term TPES indicates the energy resources in their raw form like crude oil, mined coal, etc. not the processed energy forms like petrol, diesel, electricity, etc. Since typically TPES is used to represent the energy usage of the country, which is a large number, it is normally presented in units of Megatons of oil equivalent or Mtoe. It indicates the amount of energy released from burning of 1 million tons of oil.

The use of coal started in the 1850s, the use of oil started in 1870s and the use of gas started in 1890s. Ever since their first use, the amount of usage of each of these fossil fuels has been increasing.

In 2016, the total TPES of the world was more than 13700 Mtoe, in which the share of fossil fuels (coal, oil and gas) was more than 81%*. It shows how the world energy consumption is dominated by fossil fuels. The use of fossil fuel in the world in the 1970s was in the range of about 85%. Though the use of renewable sources is increasing but only by small margins.

In the last 50 years, global energy demand has tripled mainly due to the usage pattern, population growth, economic development, and innovations in technology. It is projected to triple again over the next 30 years.

The same scenario is also true for most of the large economies of the world which include the USA, UK, Japan, Germany, France, Canada, Italy and Australia. The high GDP of the large economies is fuelled by the use of fossil fuels. For example, the contribution of fossil fuel in TPES of the USA and the UK is over 80%, for Canada, it's over 74% and for Japan, it's over 90%.

The scenario for most of the developing countries or small economies is also not very different either.

Electricity is an important form of energy driving the growth of the modern world and it continues to position itself as the "fuel of the future," as the form of fuel is changing from solid to liquid to gas to hopefully to electricity. In 2018 more than 23,000 billion units of electricity was generated in the world. This rapid growth is pushing electricity towards a 20% share in the total final consumption of energy. About 65% of total electricity in the world is produced using fossil fuels.**

GENERATING ELECTRICITY FROM FOSSIL FUELS IS DONKEY'S WORK

Since fossil fuels are mainly concentrated in a few countries, most other countries rely on the import of coal, oil and gas from other countries. In this context, let us have a look at how electricity is generated and consumed using fossil fuel, let's say from coal.

Let us take an example of India. India imports nearly 200 million tons of coal every year, a portion of it is imported from as far away as Australia.

"*How is coal imported?*" I normally ask in my seminars.

"*By the ship,*" most people immediately answer.

"*But the coal is not produced at the harbour of Australia itself.*"

I throw the discussion back to their side.

Coal is mined somewhere inside the land in Australia, then the trains and trucks are used to transport it to the harbour of Australia, then it gets loaded into the ship, which then travels a distance of on an average of about 8,000 to 10,000 kilometres to come to the harbour of India. Here, the coal is then loaded again in the trains and trucks to get transported to power plants for electricity generation. At the power plants, the coal is burnt to generate steam. Most of the energy from the coal is lost at this point as the typical power plant efficiency is only about 35%. It means that almost 65 % of the energy contained in the coal gets lost.

I hope the readers know that it takes millions of years for nature to convert dead plant matter into coal. And, 65% of this nature's work of millions of years becomes useless in a matter of minutes and hours during combustion of coal in power plants. Wow! Humans, very intelligent!

Let us continue our discussion on how overall electricity is generated and utilized using coal. After travelling thousands of kilometres, the coal is burnt to generate heat. The heat derived from the burning of the coal is used to generate steam, which is then used to run the steam turbine and obtaining rotational motion. This rotational motion is then used to run electricity generator through which very high voltage electricity is generated.

Since the users of electricity are not located where the power plants are, therefore, the generated electricity is then again transmitted to hundreds and thousands of kilometres at very high voltage levels, in the range of 600,000 Volts or 600 kilo-Volts or 600 kV. In technical terms 1000 is represented as 1 kilo or simply by symbol 'k', hence 600,000 Volts is 600 kV in short. The actual value of this voltage level varies from country to country. The transmission cables of large diameters are used, which are very heavy and require very heavy pole structure to mount and hold them.

Such high voltage levels cannot be utilised directly, therefore the next step is to step-down or reduce the voltage levels using transformers. The 600 kV is stepped down to 400 kV this high voltage is again transmitted to the neighbouring areas probably hundreds of kilometres and it once again gets stepped down to 220 kV using transformers. This is still a high voltage and it further gets transmitted to another substation where it gets stepped down using another set of transformers to 110 kV, then to 33 kV, then to 11 kV.

Finally one last time the voltage gets stepped down to 230 Volt for domestic applications or 440 Volt for industrial applications (this final voltage level varies from country to country). And, finally, this electricity passes through the electricity meters before we use it to run our appliances.

Can the readers believe it, that actually we generate and use electricity in this manner? On top of it, we waste electricity. We forget to switch-off appliances or we keep these switched on even when not in use. Isn't this wastage criminal? After reading this book, hopefully, the readers will be more sensitive towards the use of coal-based electricity, especially after knowing that it has reached them by travelling hundreds and thousands of kilometres.

So you know the quantum of infrastructure that gets utilized for electricity generation and supply to our houses or works. The most unfortunate aspect, however, is the entire efficiency of the process. From the mining of coal to electricity consumption at the last point, the entire conversion efficiency comes to only about 8 to 10 per cent. Isn't it an overwhelming utilisation of resources? Isn't it a very complex arrangement, above all, isn't it a donkey's work?

This is just an example of using coal for electricity generation but similarly, long path and various conversion steps are taken for exploiting other fossil fuels like crude oil and natural gas as well. Similarly, the inefficiencies are high, the infrastructure is complicated, and the environmental impacts are tremendous. In fact, this last aspect of environmental damage due to the use of fossil fuel is not fully understood by humans.

NO ONE WANTS COAL OR NUCLEAR POWER PLANT

We all want to make use of energy but do not want a nuclear or coal power plant in the backyard!

Unfortunately, in most countries, the major portion of electricity is generated using a mix of coal, oil, gas, nuclear and hydro. For example, in most countries and in the major electricity-producing countries like the US, China and India, more than 65% of electricity is generated using fossil fuels. Electricity is becoming one of the major final forms of energy that we use to perform several tasks of our day-to-day life at home as well as in factories, offices and in agriculture.

In various forums I have been asking this question, *"If I have a beautiful coal power plant technology that is good enough to generate electricity to fulfil all your needs, and there will be an arrangement to supply you coal every single day like your milk supply, would you like to install coal power plant in your backyard?*

Nobody says 'yes'.

I have been asking the same question for the installation of nuclear power plants in the backyard of the house, the answer is 'no'. But we all want to use electricity. There is an apparent threat from nuclear power plants and no one wants to take the risk. There is pollution from the coal power plant and no one wants to pollute their own backyard. But we all are fine if the threat of nuclear plant is somewhere else, we all are fine if the pollution caused by coal power plants is somewhere else. Isn't it? Is this not ironic of humans to behave like this? Is this not the double standards, that we humans, any way adopt in so many other aspects of life as well?

IS FOSSIL FUEL USE SUSTAINABLE?

Is such rampant use of fossil fuels to run our lives and our economies sustainable? Is reliance on fossil fuels is sustainable in the long run? Fossil fuels are finite resources and will not last for a very long time. There are plenty of books describing the peaking of fossil fuel production and their decline. At current rates of production, the oil will run out in 53 years, natural gas in 54 and coal in 150 years as per CIA factbook[$].

We started using fossil fuel some 100 to 200 years ago and in 100 to 200 years in future, we are talking their end. How could we think of putting our faith in the finite energy source for potentially infinite life?

Beginning and end of fossil fuels within the visible horizon for mankind should not be shocking news for any informed reader. But, despite knowing this fact, we continue to run our lives on these fuels.

This just gives us the idea of the tremendous pace of change in the last 150 years, as compared to the change which happened in millions of years. Is this good or bad? From one angle it is good as it has resulted in tremendous industrialization, increase in the level of production, resulting in longer lifespans and bigger comfort. But from the sustainability angle, it looks terrible.

It has resulted in a decrease in forest cover, erosion of soil, rise in air and water pollution and above all, it has resulted in climate change. The last 150 years of industrialization have changed our habitat enormously. Let's us weigh which side is heavier!

Carbon emissions from the use of fossil fuels are leading to climate change. The scenario is really scary, the time has come when we start using renewable energy sources like Solar which originates from the most powerful and infinite source – The Sun. Harbinger of all life on the planet, the sun is the primary source of energy whose power will never deplete, how much so ever you use it. Data shows that countries have started showing inclination in favour of Solar energy due to various merits.

Ref : *www.iea.org
      **www.data.worldbank.org
      $www.cia.gov

CHAPTER 15

FOSSIL FUEL USAGE CAUSES CLIMATE CHANGE

Our planet Earth has two neighbouring planets – Venus and Mars. As compared to the position of the Earth with respect to the Sun, Venus is closer to the Sun and Mars is farther from the Sun. Both of these planets cannot host human lives due to their respective temperatures owing to their distance from Sun besides other factors like the amount of oxygen, availability of water, etc. which are also important for human survival.

The surface temperature of Venus is 470-degree centigrade and of Mars is 30-degree centigrade below zero, whereas the average surface temperature of the planet Earth is about 15-degree centigrade, just ideal to support life on the planet. This is all because of the amount of Solar radiation received by a planet and peculiar climate of each of the planet. A very strong greenhouse effect on Venus makes it's surface too hot and weak greenhouse effect on Mars makes it's surface too cold.

GREENHOUSE EFFECT

The greenhouse effect makes the planet's atmosphere warmer, without which the Earth's surface temperature would have been 18-degree centigrade below freezing point. While with the greenhouse effect the Earth's surface temperature is about 15-degree centigrade, which makes it suitable for the life of all kinds on the planet. Therefore, the greenhouse effect is very essential for human existence.

The composition of the atmosphere of a planet determines the amount of greenhouse effect it will have. When sunlight falls on a planet, part of the light energy gets reflected from its atmosphere and part of it penetrates the atmosphere and gets absorbed on the surface. The surface re-radiates the absorbed energy towards the sky, but the elements or gases present in the atmosphere absorb the radiated energy and heats up the atmosphere. This trapping of radiated heat from the surface of a planet by the atmosphere is called the greenhouse effect. And, the gases which cause this heating are referred to as greenhouse gases.

The primary greenhouse gases in Earth's atmosphere are water vapour, Carbon Dioxide, Methane, Nitrous Oxide, Chlorofluorocarbons and Ozone. The contribution of each of these gases in the greenhouse effect depends on their chemical property, their lifetime in the atmosphere and in the quantities that they are present in the atmosphere.

The atmosphere on the Earth has somehow got this balance right, which has been maintained for millions of years, which is the primary reason for life on Earth for millions of years. However, a little less or little more of greenhouse effect can disturb the balance and can result in a situation wherein no life can be sustained.

In the current times, human activities are resulting in the change in the composition of the Earth's atmosphere, which, in turn, is resulting in a change in the amount of greenhouse effect. As a result, the climate is changing faster than one can imagine. Let's look into details about how it is happening? Is climate really changing? Could human activities lead to a change of such magnitude that it may be the cause of human extinction from the planet Earth?

CARBON DIOXIDE (CO2) EMISSION – WORRISOME MOLECULES

In the Earth's atmosphere the non-greenhouse gases, the gases which do not cause the greenhouse effect, makes most of the atmosphere. The Earth's atmosphere consists of 78.09% Nitrogen, 20.95% Oxygen, 0.93% Argon, this itself makes it to 99.96%. Thus, almost all of the atmosphere of the Earth is made up of non-greenhouse gases. The remaining small portion of the atmosphere includes water vapours, very small quantities of other gases and greenhouse gases.

Among the various greenhouse gases, today the Carbon Dioxide is a major cause of worry as far as climate change is concerned.

For thousands of years the average concentration of Carbon Dioxide in the atmosphere was maintained by nature to about 280 ppm (parts per million), meaning in One million parts of atmosphere there were only 280 parts of Carbon Dioxide*. Carbon Dioxide gets absorbed from the atmosphere and gets emitted to the atmosphere by various processes. It is absorbed by the plants and trees by photosynthesis process, besides being absorbed in the sea surface and other water bodies.

In the opposite process, Carbon Dioxide gets released from the land and sea due to respiration and decomposition of organic matter. Both these opposite processes were in perfect balance and net concentration of Carbon Dioxide in the atmosphere was maintained to 280 ppm. These natural processes have been happening on the planet for millions of years. As a result, the concentration of Carbon Dioxide in the atmosphere fluctuates from season to season, but the average level remains nearly constant.

These days, however, the Carbon Dioxide balance is getting disturbed, with the rise in its concentration due to human activities, as a result, the percentage of Carbon Dioxide in the atmosphere is increasing. The increased concentration of Carbon Dioxide is resulting in an increased greenhouse effect, and this increase is becoming a major concern.

Main activities which are resulting in an increase in Carbon Dioxide include:

1. Most of the Carbon Dioxide is released to the atmosphere by burning of fossil fuels for electricity generation, industrial use, transportation and cooking. Fossil fuels are a source of Carbon which on burning gets converted into Carbon Dioxide. This way, the Carbon which was sitting underground in the form of coal, oil and gas, gets released in the atmosphere which leads to climate change.

2. Deforestation and food production are other major causes of the increase in Carbon Dioxide in the atmosphere. The trees and plants are good sinks of Carbon and when they are cut and burnt, the Carbon in the wood gets released as Carbon Dioxide in the atmosphere. If emissions associated with pre and post-production activities in the global food system are included, the greenhouse gas emissions are estimated to be 21-37%** of total net anthropogenic emissions, as per the recent report from the Intergovernmental Panel on Climate Change (IPCC).

PATTERNS OF CO2 CONCENTRATION

When we look at the amount of Carbon Dioxide in the atmosphere in the long-term, we realize how much change we have inflicted on the atmosphere and more importantly how fast. The graph below shows the percentage of Carbon Dioxide in the atmosphere represented in terms of parts per million or 'ppm' for the last 1000 years. There are various scientific methods by which the atmospheric concentration of Carbon Dioxide levels in the atmosphere were measured. It is found that the Carbon Dioxide concentration was 280 ppm for most of the last 1000 years and started increasing only since 1850, the time when the industrial revolution started.

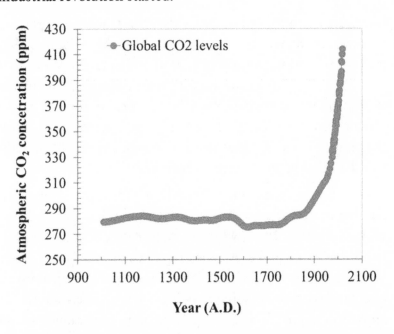

The industrial revolution is characterized by the use of fossil fuels, starting with commercial use of coal in the 1850s, the beginning of the use of crude oil in the 1870s and then natural gas around the 1880s. The increase in Carbon Dioxide in the atmosphere has a direct correlation with an increase in production and consumption (burning) of fossil fuels since the last 170 years.

Carbon Dioxide in the atmosphere has increased by 47% from its pre-industrial era level (as recorded in 2018)[#]. That is from 280 ppm in 1850 to nearly 413 ppm in 2018. Anything that changes by 47% has to be considered a major change.

If we still keep a blind eye to this fact, we are likely to be doomed.

CLIMATE HAS CHANGED – CHANGING FASTER

Recently, I have travelled across the world, across continents. In every country, I asked people, both highly educated (like professors) and not so highly educated (like taxi drivers), have they felt any change in climate within their lifetime? The answer is Yes, almost everywhere, be it in America, Europe, Asia, the Middle East or Africa. We do not even need scientific proof that the climate is changing or it has changed. People, all over the world are feeling this change in their daily lives.

I come from a farmer's family. I remember doing little farming myself at an early age, so when in the early 80s, in my village in MP in India, when we used to grow gram crop, no irrigation was required. A gram crop is a short 60 days crop and sowing happens in the winters. There used to be enough moisture in the soil from sowing to harvesting hence no irrigation for the crop was required. This is not possible now. One cannot get the gram crop without irrigation. It is evident that something has changed.

Recently in April of 2019, there was a news that the wheat crop was destroyed in two different regions of Pakistan – Punjab and Sindh – neighbouring provinces. But the astonishing fact was that in Punjab, the wheat crop was destroyed because of floods and in Sindh, it happened because of the draught. Imagine the crop getting destroyed because of two starkly different reasons in a small country.

As I am writing a book, it is summer in Europe, the temperatures in Europe are soaring to an unheard high level, one of the newspapers even reported that '*Europe is melting*'. The temperature in Paris crossed 45-degree centigrade. I lived in Europe for 5 years and it is really unbelievable and unimaginable that such high temperatures are possible in France. Events like these are predicted by scientists as an effect of climate change.

Rate of Carbon Dioxide emission is increasing; hence the rate of climate change is on the rise. It is clear from the available data that of all the man-made Carbon Dioxide emission happened in the entire history of human civilization, 50% of it came in the last 30 years from 1989 to 2018. And, the last 10 years, that is from 2009 to 2018 alone contributed to more than 20% of the total Carbon Dioxide emissions[@].

Can you feel the pace at which the damage is increasing?

The ice cap cover on the arctic pole was 8.6 million square kilometres in 1950, which has reduced and in 2015 it was only 4.6 million square kilometres, a decrease of more than 45%[$]. The ice cover is decreasing at a very fast rate, thus disturbing the ecological balance in polar regions. Also, reduced reflection due to reduced ice coverage is resulting in less reflection and higher trapping of energy adding to an increase in climate change.

How anyone on the Earth can imagine a bright future for themselves and their near and dear ones in this scenario?

IPCC REPORT – GLOBAL WARMING OF 1.5 - DEGREE C

The IPCC or Intergovernmental Panel on Climate Change is an Intergovernmental body of the United Nations dedicated to providing the world with an objective and scientific view of climate change. The panel also gives input on natural, political and economic impacts and risks of climate change as well as possible response options.

In response to a request from the United Nations during the Paris Agreement of 2015, the IPCC was given a task to prepare a report on the impacts of global warming of 1.5 - degree C above pre-industrial levels and related global greenhouse gas emission pathways.

This report titled "Global Warming of 1.5 - degree centigrade" was published in 2018. The report has come in the context of strengthening the global response to the threat of climate change, sustainable development, and efforts to eradicate poverty. The report brings forward the impact of global warming on ecosystems and economies, and more importantly, how it enhances the problems of disadvantaged. The report also highlights the differences in the scenario when global warming is limited to 1.5 and 2-degree centigrade.

The IPCC[&] report summarizes that:

* The world is already hotter by about 1-degree centigrade as compared to average global temperatures of the pre-industrial era

* Limiting global warming to 1.5 - degree C would require *rapid and far-reaching transitions in energy*, land, urban and infrastructure (including transport and buildings), and industrial.

* Limiting global warming to 1.5 - degree C requires a reduction in emissions of Carbon Dioxide due to human activities to 45% from 2010 levels by 2030 and it should *reach net Carbon zero around 2050.*

As per the IPCC recommendation, is there any region or country in the world which is making *'rapid and far-reaching transition in energy'*? Actually None!

During my travels around the world in the year 2019, I found that no one seems to be taking this very seriously. Also, as per IPCC, all of us should become net Carbon Zero by 2050. It means that if we are not able to extract any Carbon Dioxide from the atmosphere, we should not emit any Carbon Dioxide.

But any use of fossil fuels results in Carbon Dioxide emission, which means that we need to stop using fossil fuels by 2050. Can we imagine a world by 2050, wherein there will be absolutely no usage of Carbon in any form be it coal, petrol, diesel and LPG?

It is really unimaginable! Is it not?

Though there are efforts going on in the world to find out ways and means to extract Carbon Dioxide from the atmosphere so that we can reverse climate change. However, so far, no such technology has been developed which can do this in a reasonably cost-effective manner and chances are bleak that we develop such a technology. And, even, if we are able to develop such a technology, ever-increasing consumption of resources by humans would make life unsustainable on the planet.

IMPACT OF CLIMATE CHANGE ON OUR LIVES

Limiting global warming to 1.5 -degree centigrade is really ideal in the current context. The ultimate limit to global warming is

2- degree centigrade, beyond which the effect of global warming will be much more catastrophic and even result in irreversible change, implying that even with all possible corrective measures, we will never be able to restore our climate to its original status.

How terrible it would be?

Climate change means an imbalance in a natural ecosystem, in which we are living. This imbalance is resulting in changes that are already affecting our lives and will affect more severely in the future. Needless to say, that the effect on lives will be much higher in the case of 2-degree centigrade global warming as compared to 1.5-degree centigrade global warming. The average global warming looks small, just 1.5 to 2-degree centigrade, but its impact is going to be enormous on our lives as predicted by thousands of scientists around the world.

Here are some of the expected impacts of climate change on our lives, in case of global warming by 1.5 to 2-degree centigrade:

- Climate change will cause heatwaves, longer than the normal duration. If the global temperature rises by 1.5-degree centigrade, the heatwave may be of longer duration - up to a month or more.
- The shortage of freshwater will be in the range of 10 to 15%.
- The intensity of drought and rainfall will increase.
- Production of the yield of all basic crops - wheat, maize, rice will decline, 10 to 15% in case of wheat.
- Sea level will rise by 40 to 50 centimetres, which would force millions of people living in the low-level coastal areas to migrate, adding more strain on the resources of locations where they migrate.

- Higher temperatures would lead to a faster spread of diseases.

- The use of finite fossil fuel is the cause of concern. The size of fossil fuel reserves and the dilemma that "when fossil fuel energy will be diminished" is a fundamental question that is disturbing the thinkers of this world. But why this question? How do we view fossil fuels and how should we view fossil fuels, needs to be understood, which needs to be changed, for the benefit of humanity?

LIKE DINOSAURS, WILL HUMANS ALSO EXTINCT FROM THE PLANET?

In the business as usual scenario, meaning that if we do not take corrective measures that are required to limit global warming, and we continue to live the life in the manner that we are living right now, the rise in global temperature by the end of the century i.e. by 2100 is estimated to be about 4 to 5 degree centigrade.

If and when the temperature rises to such a level, it would put a question mark on the existence of human beings. The changes in the climate will be very drastic that a non-zero finite possibility of human extinction exists. There is a big question mark on human existence. As we say today, that once upon a time there were dinosaurs on this planet, some other life form, that may come on this planet thousands of years later, they will say that once upon a time there were humans on this planet. How terrible it is that you are reading these lines! The possibility of saying that *"Once upon a time there were humans on this planet."*

Since energy generation and consumption are major culprits resulting in climate change, should we change our lifestyle completely, so that our grandchildren and their children will have the possibility to live on this planet, we call our home? Are we so consumed in our own life and exploiting nature so much that we cannot even think for the good of our own grandchildren? Should we all shift to use of energy sources that are natural, renewable and limitless? Should not the limitless possibility of human life on this planet use the limitless energy resources?

The Solar energy!

We need to take full U-turn on energy generation and consumption, we need to stop using fossil fuel and switch to renewable energy, completely and immediately.

I would quote the words of Saudi Oil Minister Sheikh Ahmed Zaki Yamani in 2002 here, *"The Stone Age didn't end for lack of stone, and the oil age will end long before the world runs out of oil."*

As Sheikh Ahmed envisaged, the use of fossil fuel must stop. We need to stop the damage being caused to our environment for the continuous and long-term sustainability of life on the planet.

We need to switch to sustainable energy to sustain life on the planet.

Ref : *Carbon Dioxide Information Analysis Center
    (CDIAC) and the World Resource Institute and
    Carbon Brief
    **www.iefworld.org / www.ipcc.ch
    #www.climate.nasa.gov
    @www.iea.org
    $www. earthobservatory.nasa.gov
    &www.ipcc.ch

CHAPTER 16

SOLAR ENERGY FOR SUSTENANCE

Solar energy has been nurturing the lives of millions of species on the planet for millions of years. Unlike fossil fuels, Solar energy is abundantly available. The planet receives 10,000 times more energy in the form of sunlight than the annual consumption of entire humanity in all possible forms. As argued earlier, the human existence on this planet has been there for a long time and we should design our energy generation and consumption system based on the comparatively infinite energy source, the Sun, and definitely not on the very finite fossil fuel-based energy sources.

SOLAR ENERGY GETS MANIFESTED IN ALL ENERGY FORMS

It is worthwhile to note that Solar energy gets manifested in many other forms on the planet. We know that it is because of the Solar energy that the planet has warmth which is required for sustaining lives. It is because of the Solar energy, the photosynthesis process happens in the plants, which not only helps us to provide Oxygen for a living but also food for our survival. The growth of plants and trees also provides us energy. All this is manifestation of Solar energy.

The shape of the Earth, its inclination on its axis and uneven geography, results in uneven heating of its surface due to sunlight. This uneven heating of the Earth's surface results in the blowing of the Wind, which is a form of energy used for sailing, drying clothes, etc. But in the current times, the wind energy is also used for generating electricity, hence in Wind energy, there is a manifestation of Solar energy.

The evaporation of water due to heat, which in turn is provided by Solar energy and its condensation results in rains on the planet. When it rains or snow falls, there is a flow of water on the mountain and in the rivers. The flowing water directly or stored water in the dams is utilized to generate electricity. Thus, Hydro energy is also a manifestation of Solar energy.

Fossil fuels are remnants of plants and animals, processed over millions of years by nature, hence a manifestation of Solar energy. Overall, it is Solar energy that nurtures the planet and gets manifested in many ways, which we can use and convert into a useful form for fulfilling our domestic as well as industrial needs.

Solar energy as sunlight or in other forms is available in abundance. But the main question here is how to convert it into a useful form which we can use to perform our tasks. In the following sections, we will discuss how we can convert Solar energy into useful energy at a conceptual level and the possibility of generating and fulfilling all our energy needs using Solar energy.

Sunlight can be directly utilized to fulfil almost all our energy needs. There are two main technologies developed and utilized by many people around the world; (1) Solar cell technologies wherein the sunlight gets converted into electricity, and (2) Solar thermal technologies wherein the sunlight gets converted into heat energy.

SOLAR CELL TECHNOLOGIES

A Solar cell is a building block of Solar cell technologies. When sunlight falls on a Solar cell, it gets converted directly into electricity without any parts moving in it. Compare this, with coal power plant for generation of electricity, the amount of infrastructure required to generate electricity using coal, its mining, its transportation, its combustion, then the generation of electricity, its transmission and its distribution to users.

On the other hand, the Solar cell converts incoming sunlight to electricity, without any of the above steps, without moving any parts, without any major wear and tear. A Solar cell can generate electricity instantaneously, anywhere in the world. Knowing the process of coal electricity generation, it is almost magical that Solar cell generates electricity in such a simple manner, isn't it?

Technically, the generation of electricity by Solar cell due to sunlight falling on it is referred to as Photo-Voltaic (PV) effect. Therefore, sometimes Solar cells are also known as PV cells.

Materials that can be used to make Solar cells are referred to as semiconductors. A Solar cell consists of two electrodes or two terminals, as in the case of a battery, a positive and a negative terminal. These two terminals in scientific terms are known as P-type semiconductors and N-type semiconductors. A Solar cell is a junction between these two semiconductors and it is also referred to as P-N junction Solar cells.

When sunlight falls on a Solar cells, the P & N terminals get charged and drive the current if connected across an appliance, provided a sufficient amount of current and voltage is generated. One of the beautiful aspects of Solar cell technologies is that the cells can be connected with each other, in a serial manner or in a parallel manner to generate literally any amount of current or any amount of voltage that we want.

I guess readers would not mind these little technical terms, as I believe that these Solar cells could become a big saviour for humankind. We can generate electricity using these Solar cells and as I described earlier, with the use of electricity, we can perform any task in the world and fulfil all our energy needs for cooking, lighting, heating and transportation etc.

SOLAR CELL TECHNOLOGY CAN FULFIL ALL OUR ENERGY NEEDS

Once we interconnect the Solar cells to get higher voltage and current, it is called Solar module, or Solar PV module. Here we can use Solar cell technology for powering small appliances like calculators and watches, to larger devices like fans and TVs, to the larger applications to power our entire house, to even larger applications to power institutions and factories, to even larger application to power entire cities and countries. Yes, indeed!

Still one question that comes to our mind is, would Solar cell be able to fulfil all our energy needs? Would it be able to fulfil the energy needs of each citizen of every country?

Once again, the answer is a big 'yes'.

There are several materials that are being used to produce Solar cells. One of the workhorses of Solar cell technology is the material called 'Silicon' which is abundantly available and is the second-largest material available in the Earth's crust. It is non-toxic material and one can make Solar cells with very long life, over two decades, at least.

This way, Silicon is one of the most commonly used materials in the manufacturing of Solar cells and modules. Not only Silicon there are some other materials that are used to some extent to make Solar cells and scientists world over are working to make even cheaper and long-lasting Solar cells.

Personally, I have been dealing with Silicon for a long time. During my Ph.D. in Belgium, I did research on making thin Solar cells of Silicon to reduce its cost. And, after my Ph.D. I continued working with Silicon. Considering all the facts about Silicon and other material for Solar cells, I can certainly say that the energy needs of entire humanity, in all possible forms, can easily be fulfilled with Solar cell technologies, as it generates electricity and with electricity, we can fulfil all our needs.

SOLAR THERMAL TECHNOLOGIES

In Solar thermal technologies, the energy of the sunlight is converted into heat energy. This can be used for many applications like cooking and water heating for domestic applications, and providing heat for processing material in industries. With heat energy, one can do both cooling as well as heating. One can use this heat energy for heating of spaces, required in cold countries and for cooling of spaces, required in warm countries.

In our life, we have experienced the conversion of sunlight into heat energy. If a metal piece is kept under sunlight it becomes hot, but if a piece of wood is kept under sunlight it does not become hot. Metals are good absorber and converter of sunlight to heat energy. Two common Solar thermal technologies are Solar water heater and Solar box type cooker. In both these technologies, water gets heated, and in one case hot water is used for bathing and in another case, it is used for cooking.

We all must have played with a magnifying glass in our school time, not for doing the experiments of physics, but for the fun of burning the paper. The magnifying glass collects the light from a large area and focuses it onto a small area, thereby generating high temperature in the small area, which burns the paper. The same principle is used in Solar thermal technologies for generating higher temperatures. In fact, Solar thermal technologies are very effective in generating a range of temperatures right from 60 to

70-degree centigrade in water heaters to as high as 1000 degree centigrade for industrial applications.

Not only that, the high-temperature heat can also be used to generate steam and like in a coal power plant, it can be used to generate electricity. In this way, the Solar thermal technologies can be used for a wide range of applications.

There are several limitations of Solar thermal technologies due to which their use is not growing the way use of Solar cell technologies is growing. For generating higher temperatures, Solar thermal technologies need to continuously follow Sun. This makes operations cumbersome, requires continuous tracking of the Sun by way of movement of parts, hence a good amount of maintenance.

This way we see that every energy demand can be met by generation of Solar energy. Today there are examples where people have started using Solar energy for lighting and running fans, refrigerators, ACs and even cars. There are even several examples where the entire house, institution or even a factory are fully running on Solar energy.

IGNORANCE ABOUT SOLAR ENERGY

During my several seminars, I have been asking people *"How many of you use Solar energy in your life?"*

Not many people raise their hands. Only about 5 to 10 per cent of people raise their hands.

This shows the ignorance of people. I ask a few more questions: *"How many of you have been eating food? Is the consumption of food, not the use of Solar energy? Is the breathing of air is not the use of Solar energy? Is the use of water is not the use of Solar energy?*

Then people realize that all of us are big users of Solar energy.

By using fossil fuel energy to a great extent, we have made fossil energy as centre of our lives. We have forgotten that the Sun is the centre of our life, the very reason for our existence, without the energy of the Sun, the planet will not be warm enough for our comfort, we will not get food, air or water either. The human life can exist for millions of years only when we synchronize our lives with nature, only when we fully generate and fulfil ALL our energy needs using Solar energy, without affecting the climate.

It's ONLY usage of Solar energy which can make life sustainable on the planet.

We need to bring the Sun back to the centre of our lives.

PART - F

IDEA OF ENERGY SWARAJ – MY WORLD ENERGY VISION

Not mass production, but production by masses.

- Mahatma Gandhi

CHAPTER 17

WHAT COULD BE THE MODEL FOR ENERGY SUSTAINABILITY?

Having discussed the use of energy as an integral part of life, the threat of climate change due to use of energy, the finiteness of resources on the planet, the ever-growing consumption and population, the conditions for the possibility of human existence and the use of Solar energy solutions for sustenance, now it is time to discuss the best possible model for energy generation and consumption that can sustain human life on the planet in as much foreseeable future as it existed in the past, at least.

While searching for a model for energy sustainability, one has to look at centralized vs decentralized system, fossil vs renewable energy systems, grid-connected vs off-grid energy systems, government-owned vs private-owned energy systems, etc.

While searching the model for sustainable energy systems for the entire world, I have closely looked at the Gandhian Model of "Gram Swaraj" as it holds, in my view, the key to sustainable energy solutions for the world.

Gram Swaraj prominently features in Gandhian thinking: Gram means village and Swaraj means self-governance; hence Gram Swaraj is self-governance of villages. His vision was an important tool for the economic development of independent India. Gandhiji's Gram Swaraj was not about the reconstruction of the old village but creating an ecosystem where every villager had work, who also earned for the fellow villagers, besides his own family. Thus, making independent units of villages by creating a self-sufficient economy.

Mahatma Gandhi – India's Father of Nation - A great *Karma Yogi,* even a greater social and economic reformer whose whole life was devoted to the supremacy of the moral laws - truth and non-violence. He left an indelible impact on the social, economic and political spheres. He was a man of path-breaking ideas even then, encompassing all spheres in the life and times of our nation.

He used the word – Swaraj, the Republic to describe and interpret the social and economic fabric of the country, in particular, villages as India is a country of villages.

I quote him here: *"The word Swaraj is a sacred word, a Vedic word, meaning self-rule and self-restraint."*

It was the right ordering of the various powers of the self. Gandhi's ideas on the social and economic structure of the country were encapsulated in the term Gram Swaraj, which was indeed ahead of its time. Hence it needs to be understood and viewed within the context of the twin beacons of truth and non-violence.

SO, WHAT KIND OF GRAM SWARAJ GANDHI JI WANTED?

With industrialisation beginning and firming in Great Britain, it was natural that the British wanted to create centralised, industrialised and mechanised modes of production in India. Mahatma Gandhi while struggling for *Purna* (complete) Swaraj, turned this principle upside down as it envisioned a decentralised, homegrown, hand-crafted mode of production, thus propagated: *"Not mass production, but production by the masses."*

He believed in de-centralisation of power and was in favour of autonomy to villages in all respects, as he believed that this model would be more sustainable – Socially & Economically!

This could be made possible by making every village its own Republic, *"Independent of its neighbours for its own vital wants and yet interdependent for many others in which dependence is necessary."*

Each village should be basically self-reliant, making provision for all necessities of life like food, clothing, clean water, sanitation, housing, education and so on. This would include governance and self-defence, and all socially useful amenities required by a community like a theatre and a public hall. For India as a whole, full independence would mean that every village would be a Republic with full powers.

Gram Swaraj, as envisioned by Gandhiji was building an independent village, where everybody had work, hence contributed towards socio-economic growth of the village. The villagers would produce and consume the products made in their own village by their own hands for its own citizens.

A village should produce its own grains, vegetables, fruits and its own Khadi. Not only this, the village should have all those facilities which helped citizens in their day-to-day lives like schools, hospitals and theatres etc. This comprised the blueprint of "Gram Swaraj" where self-reliance would ring in by undertaking production and consumption of goods of all daily needs.

Indeed, the thought was ahead of its time. Further, Gandhiji strongly advocated for decentralization of economic and political power through the organization of Village Panchayats. Gandhiji wanted true democracy in India. He believed that true democracy cannot be worked by twenty men sitting at the centre, it has to be worked from below by the people of every village. In simpler words, there is a need to understand the fundamental concept of Gram Swaraj that every village should be its own Republic and it should work from the bottom upwards as fundamentally independence begins at the bottom.

Thus, every village will be a Republic or Panchayat having full powers. This was the picture of Gram Swaraj conceived by Gandhiji.

DECENTRALIZATION FOR EQUITY

Gandhiji's Swaraj as interpreted by German Statistician E.F. Schumacher:

'Man is small and man is - or ought to be - beautiful and as such only the human scale economy of Gandhi's dream is appropriate.'

In the words of Schumacher: The greater the size of the production unit, the greater the separation of production from consumption. Reuniting production and consumption units was only possible if production units were small. It would be easy to manage and adaptable to local conditions. One of the enormous advantages of small-scale production, reunited with small-scale consumption, was the minimisation of transport. As increased transport only added to the cost and not to the real value of goods.

In a way, this aimed at addressing the issue of inequity by bringing production and consumption together, as it will be the greatest tool in making the world more equal.

Looking back at the economic model that we have adopted since the industrialization, and history of economic development of the last 150 years, it tells us that the world has become more and more unequal with every passing year. This inequity in a way is the cause of many problems of the world, including degradation of air, water, soil, forest, besides increasing violence and even terrorism. I strongly believe that climate change is the result of the current economic model.

Village Swaraj is man-centred non-exploiting decentralized, simple village economy providing for full employment to each one of its citizens on the basis of voluntary co-operation and working for achieving self-sufficiency in its basic requirements of food, clothing and other necessities of life.

I have been reading and hearing a lot about sustainability, from the best minds in the world. But in my view, the best formula for sustainability was given by Mahatma Gandhi, the Gram Swaraj. If every nation in the world can establish Gram Swaraj in its true sense, the world would be sustainable.

I have been thinking since long, *"Could we apply this model of Gram Swaraj to Energy?"* *"Can we generate and fulfil all our energy needs within villages, houses, institutions, towns? Can we also think of "Energy Swaraj" on the lines of Gram Swaraj?*

And, Energy Swaraj within Gram Swaraj will be the ultimate sustainable model. Nature would love it.

CHAPTER 18

THE IDEA OF ENERGY SWARAJ: A MODEL OF LOCALIZED ENERGY SELF-SUFFICIENCY

Today the world is standing at the crossroads as far as Energy accessibility and climate change is concerned. There are about 1 billion people who do not have access to electricity and then there are other billions whose excessive energy usage is causing climate change.

Can any contradiction be bigger than this, the contradiction of energy?

Today solving the puzzle of extremes of energy access and climate change together, is one of the biggest challenges for humanity.

Is it a challenge? Is it a puzzle? Is it an opportunity?

This is a mixed bag of all...

It is a challenge to work towards changing the current scenario, changing production and consumption habits, accepting the fact that Earth is under tremendous pressure. It is a challenge for everyone to accept these facts.

It is a puzzle – yes, we are in a Catch 22 situation, if we give up modern energy usage, the development, growth, socio-economic standing will be hit. If we don't then we are living in the fear of drastic climate change.

Opportunity – yes for changing our prevalent habits where there is rampant usage of energy. We don't move an inch without using energy, it is the most important component of the modern lifestyle, growing every minute with the growing needs and wants, evident from the increasing inequity.

Access to energy has been proven to be a great enabler of social and economic growth in the world and affects all social and economic development indicators in a major way. Energy generation and consumption in the world is a very complex system. It is limited by or enabled by the availability of resources, geography, economics, the density of population and status of industrial growth.

This complex energy system has resulted in the unequal opportunity to communities in different parts of the world, therefore unequal growth. Climate change, energy security, air and water pollution and limited availability of fossil fuels are other concerns of current energy solutions.

Then, in this context, *what is the way we should develop our energy solutions? How should our new energy solutions look like?*

I believe that we should develop our energy solutions in a manner that:

a. provide quality and sufficient energy access to all,

b. does not affect the climate in a negative manner, and

c. provides a sustainable energy supply in the long-run.

This chapter presents an alternative model for energy generation and consumption that would fulfil these requirements of energy solutions mentioned above.

THE IDEA OF ENERGY SWARAJ

Finding sustainable energy solutions requires a clear view of the consumption patterns of humans on Earth. Consumption of any commodity, including energy, has to be in sync with nature, meaning generating and consuming only which is renewable. Else, sooner or later, the existence of human life on the planet will negatively affect the climate. Once, the sustainability of the energy solution is ensured, the new energy model should ensure access to sufficient quantity and quality of energy to enable social and economic growth. Access to energy, with equal opportunity, can bring greater equity in the world.

In such a context of climate change, energy access and need for sustainable energy systems, a model of Energy Swaraj is presented.

Modelled on the lines of Gram Swaraj, I propose Energy Swaraj through this book, where communities and individuals can opt for energy independence. They will generate and consume their own energy produced through renewable energy like Solar energy, Wind energy, Biomass Energy or combinations of these sources.

Energy Swaraj is envisaged to be a model in which communities across the world are self-governed in terms of generating and fulfilling their energy needs using locally available resources.

Like in Gram Swaraj, in Energy Swaraj, an ecosystem is to be built in such a manner that generation, installation, maintenance, manufacturing (as much as possible) and consumption of energy, mainly electricity, is done by the locals for the locals. Energy Swaraj can be defined as "Localized Energy Self-sufficiency."

In this way, the local communities can become independent in energy and can self-govern their energy production and consumption. In Energy Swaraj, it is envisaged that every household, institution, community, region and country would generate and fulfil their own energy needs.

In order to completely understand and absorb the idea of Energy Swaraj, let me lay out the idea of localized energy self-sufficiency differently:

* Energy Swaraj is a state of self-regulation where individuals and societies become sensitive to energy needs, they govern the generation and consumption of energy from within their locality using locally available resources.

* Establishing Energy Swaraj means generating and consuming energy in a holistic manner, without affecting the environment and the sustainability of life on the Earth.

* A system of generation and consumption of energy by the locals, for the locals – Energy Swaraj is a model of democratization of energy.

* Energy Swaraj is about generating and fulfilling ALL energy needs on a 24x7 basis, but before generating energy, avoiding the energy needs if avoidable, minimizing the energy needs, if minimizable and generating energy only if it is really required.

This would lead to many benefits for the communities including independence in energy, energy sustainability, job creation, confidence in technology, skill development, women empowerment, that too without any negative impact on the climate. Hence, such energy ecosystem could bring great opportunity to local communities which will enable social and economic growth with equal opportunity.

The next question is whether such a localized system of energy generation and consumption is viable at all? This is discussed in the next chapter.

IS ANY AMOUNT OF RENEWABLE ENERGY GENERATION SUSTAINABLE?

Is the mere generation of all the energy needs using renewable energy sources could be good enough condition for sustainability? I feel that the finer aspects of energy generation and consumption, even if it is from renewable energy, is very important to understand. Suppose we have a magic wand using which, today itself, we convert our entire energy generation and consumption based on renewable energy, in that case, would our energy solutions become sustainable?

I am afraid, the answer would be 'No'.

Generating any amount of energy, even using renewable energy sources will not be true to the meaning of sustainability, that is *"Development that meets the needs of the present without compromising the ability of future generations to meet their own needs."*

Production of equipment (like a Solar panel, wind turbine, etc.) and components (like structures, wires, etc.) of renewable energy solutions would not be having Zero-Carbon footprint. The Carbon footprint of renewable energy solutions would definitely be much less than the generation of energy using fossil fuels, but it will not be Zero.

From this perspective, generating a large amount of energy, even if it is through a renewable source of energy, is not going to be sustainable. It is not that because the Sun's energy will be in short supply or Sun will be shining less in future, but simply because if we do not put any constraint on the amount of consumption, the world will not be sustainable. As it has been argued in an earlier chapter that ever-increasing production and consumption of goods and services cannot lead to sustainability. This is an important aspect of sustainability and must be taken care of.

THE GANDHIAN IDEA OF SUSTAINABILITY

Gandhiji believed in de-centralisation of power and was in favour of the autonomy of villages in all respects. This model would provide long- term sustainability as a society, community or village is fulfilling all its needs internally.

The fundamental idea in Gandhian Sustainability model is "constrained consumption," which is driven by the 'need' and not by the 'greed' of the people. The quote of Gandhiji makes it very clear, he said, *Earth provides enough to satisfy every man's need, but not for every man's greed."*

Gandhiji thought of fulfilling all necessities of life from local resources, which cannot be available in infinite quantities, until and unless communities consume resources as per their needs, otherwise there is no way such a localized ecosystem can work and satisfy their demand on resources.

In the model of Energy Swaraj as well, practising constraint in consumption is key to the localized energy generation and consumption.

MY IDEAL CANVAS FOR SOLAR SOLUTIONS...

My ideal canvas of Solar solutions would include: A reasonably large geographical area to work on, say a district, where there will be several small complementary manufacturing units for Solar panels, electronic and electrical circuits, lighting devices and various other parts of energy solutions. With the current status of technologies, this can be done at a local level, without requiring very heavy investment or even sophisticated technology.

My ideal canvas would include: Several Solar shops where the shop owner will have the skills to assemble and repair the Solar products, install Solar solutions on rooftops and on the ground. The shop owners would have the capability of providing Solar solutions within the locality of their shops, let's say covering 10 to 15 villages. I believe the shop owner would be able to generate employment for a few others.

Here, the local academic institutions can skill people in their district on the technological solutions required in that area. Not only skilling, but they would also be able to continuously improve the Solar solutions, either by themselves or by studying what is happening around the world. This will keep Solar solutions or in general, renewable energy solutions upgraded all the time.

My ideal canvas would include: A community-based banking system, like societies or cooperative banks, that would provide low-cost finance to various stakeholders in the ecosystem, the manufacturing units, the shop-keepers and the users of Solar solutions.

My ideal canvas would include: Envisaging governments around the world to play an effective role by providing initial funding to the cooperative banks for this purpose. Government through the local banks will be providing loans at a low-interest rate. For the economically weaker households, the Government can provide Solar solutions through these localized models, which will save it from the hassle of large-scale procurement and it's monitoring, etc.

My ideal canvas would include: Scientific institutions across the world to keep working on higher-level science for societal reasons. This would lead to cost reduction, efficiency improvement, new materials and design, etc. which through the local academic institutions can percolate down to the society.

ENERGY SWARAJ FOR ALL – ACROSS THE WORLD

The idea and concept of Energy Swaraj encompass one and all: Rich and Poor, Have's & Have Nots, People having or not having grid connectivity and a citizen of a developed, developing or underdeveloped country. The concept of Energy Swaraj has applications in every community and everywhere in the world.

In fact, everyone across the world has to take part in it, has to adopt Energy Swaraj, as the climate has no boundaries and any act of yours, be as an individual or as a nation is affecting others.

Let me clarify here that this doesn't imply that every individual will start installing Solar panels, generating electricity, transmitting and maintaining it. There has to be a kind of ecosystem wherein a group of people are involved in the process. It would be a Solar energy ecosystem by the Locals for the Locals. In Energy Swaraj, everybody will have a specific role with regard to installation, generation, transmission and maintenance.

This is the idea of Energy Swaraj.

To sum it up, Energy Swaraj is a state of self-regulation where individuals and societies become sensitive to energy needs and govern the generation and consumption of energy from within their locality in a manner that it does not affect the sustainability of life on earth, like Mahatma Gandhi's Gram Swaraj.

175

LET'S TAKE A U-TURN

If we replace the usage of fossil fuels by harnessing the power of all-pervasive SUN in the form of Solar Energy or renewable energy, which can be generated in any quantity, anywhere in the world, and as per the requirement, it will change the scenario. It is environment-friendly and sustainable too. Moreover, this will change the entire dynamics of energy generation and consumption and create self-sustainable society. If done in a community setting, it will change the dynamics of energy usage.

Time has come for considering a U-turn NOW in energy generation and consumption pattern, rather than centralized solution, it is time to move towards decentralized energy solutions, rather than using fossil fuels it is time to move towards 100% renewable energy solutions, rather than contributing to Carbon footprint, it is time to reduce it significantly or make it zero. It is time to move towards Energy Swaraj. It is time to adopt Energy Swaraj to its fullest, across the world.

This will raise the sustainability on Earth as it will be based on usage of alternative sources like Solar – the energy derived from the sun, which is an infinite source, abundantly available and hardly impacting the climate.

MY INCREDIBLE JOURNEY OF 20 YEARS!

How amazing my journey in Solar has been!

20 years back, I wished to do something for the society, and I got the idea to work in the field of Solar, 20 years later, I am writing this book on Energy Swaraj. 20 years back I had no idea about Solar energy, and its potential, whatsoever. 20 years later, I feel that Energy Swaraj is the concept, that would save the world from catastrophic climate change.

After my doctorate studies in Solar energy, I started with a small experiment on a tiny Solar lamp, which has further expanded and touched the lives of millions of people, followed by experiments with Solar shops, Solar factory, Solar cookstoves and many others.

During all these experiments, slowly but surely, I realized that Solar energy is for anything and everything, where ever we need energy. The Energy Swaraj is a concept for generating and fulfilling ALL our energy needs, not only for a household, community or country but for the entire world. The idea of Energy Swaraj is applicable to the entire world.

From an idea to provide a Solar lamp to a child for study purpose, to the idea of establishing Energy Swaraj everywhere in the world, I personally got evolved and elevated with time, from experiment to experiment my vision got broadened, the potential and possibility with Solar energy looked more and more immense and viable with every experiment.

For every country, use of Solar energy in the form of Energy Swaraj is a solution to many problems including job creation, strengthening the economy or improving equity. Eventually, in the wake of severe climate change ONLY Solar energy appears to be a form of energy that can sustain human lives on the planet.

In fact, this is the SOLAR truth. The SUN is and always will be in the centre of our lives. Temporarily we have forgotten this fact, we have become unconscious for the time being and made fossil energy as the centre of our lives. Now it is time to regain consciousness, it is time to establish the Sun as the centre of our lives, for driving our lives, once again.

Let's all wake up to this SOLAR truth.

Let's strive for Energy Swaraj.

CHAPTER 19

VIABILITY AND BENEFITS OF ENERGY SWARAJ

There are some questions: Is establishing Energy Swaraj i.e. generation and fulfilment of ALL our energy needs 24x7 locally, possible at all? Can Energy Swaraj be established in every part of the world? Is it technologically feasible? Is it economically viable? Is it environment-friendly? Is it socially acceptable? Is it institutionally doable?

I can say that the answer to all the above questions is a positive and strong 'Yes'!

& Why do I say that...

...I have been working in the field of Solar energy for more than 20 years now and have acquired rich and diverse experience in the field. I have conducted research and did projects on the technology front as well on the social front. I have published more than 100 research papers, written four books, filed many patents. I have worked with Solar industry on one side and on the other side provided Solar lamps to millions of students, created jobs for thousands of women, created hundreds of Solar shops and Solar manufacturing factory in rural areas. I have known and proven the versatility of Solar energy through my various experiments. So, you must have understood why I say that the answer to all the above questions is positive and strong 'Yes'!

AMG TO AVOID OMG!

We carry out almost all our daily activities by using energy in some way or the other. It is discussed in the earlier chapters that even simple activity of brushing teeth involves enormous use of energy in making and transportation of brush, paste, glass, tap, washbasin and even pumping water to the overhead tank. We have become so dependent on energy that every minute of our life and every square inch of space around us has a touch of energy. With this kind of usage, even though the impact of each activity may be small, but collectively it makes a big impact.

From the above arguments, it is clear that if people in the world continue to produce and consume energy in business as usual manner, even if the entire energy requirement of the world is met by Solar energy like for transportation, domestic and industrial use etc. still it will create a huge impact on the environment, thus making sustainability questionable.

We need to adopt the Energy Swaraj model, but even then, what could be the best way to use energy without having a negative or negligible impact on the surroundings or environment? What could be a sustainable way of generating and consuming energy? Is there any way in which the energy needs of billions of people can be fulfilled sustainably without affecting the environment in a major way?

To provide answers to the above questions and a viable solution to the prevalent scenario, I have come up with a principle which I have named – AMG.

AMG stands for: 'A' to Avoid the use of energy, 'M' to Minimize the use of energy and 'G' to Generate the required energy.

Beginning with 'A' i.e. Avoid the use of energy as much as possible, even if it is Solar or renewable energy. Does it mean that you need to live in the dark or discomfort? Does it mean that you need to go back to the Stone age? The answer is No!

As I say this, I mean we should avoid active use of energy. So, what is the active use of energy? When machines are used to perform a function using energy, then it is active use of energy. One can obviously use Solar energy as much as one wants in a passive manner i.e. where machines are not used to perform the task. One can dry as many clothes as he or she wants in sunlight. Using sunlight during the day time, one can illuminate spaces in the buildings, homes and factories without affecting the environment.

There are many well-known techniques, which are known as Solar-passive techniques or Solar-passive architecture with which one can use Solar energy for natural lighting, natural ventilation, natural cooling and natural heating during the day time.

This approach of 'Avoid' can reduce your energy requirement to a great extent. Depending on the context, energy saving can be in the range from 20% to even 80%.

We need to practise discipline towards the consumption of goods and services of any kind, which is one of the first steps towards sustainability. Without this, one can never imagine a sustainable world, even if we start using 100% energy derived from renewable sources.

As a choice, with consciousness, at my home in Mumbai, we have stopped using the refrigerator, washing machine, geyser and microwave. Now, neither we are using them nor we are missing them.

Let's talk of 'M' of the AMG principle. We understand that beyond a certain point, you cannot avoid the use of energy in daily activities, so you need to follow 'M' in energy consumption. 'Minimize' the use of energy as much as you can by using high-efficiency or super-efficient appliances with lower energy consumption. The Light Emitting Diode (LED) is one of the best examples of an efficient device. The energy requirement of an LED is 10 times less than the incandescent lamp for the same amount of light output.

Similar to LED, there are efficient fans, TVs, refrigerators, ACs and many other appliances. These have significantly lower energy consumption than normal appliances. The efficient appliances may be expensive in the beginning, but lower energy consumption (hence lower electricity bill) during their lifetime makes up for the initial higher investment.

This approach of minimizing the use of energy further reduces our energy requirement significantly. If we follow this strategy, it can further result in energy saving in the range of 10% to 50%. With the combination of Avoid and Minimise, one can cut down on energy needs by as much as 30 to 80%.

If you have tried your best to Avoid and to Minimise, but you cannot do it beyond a point, the last option should be the 'G' of AMG principle, that is Generate energy.

But it should absolutely be the last option. In the context of Energy Swaraj, if we follow AMG, we can meet all our energy requirements through localized energy generation by means of renewable energy or Solar energy.

And, if people do not follow this principle of AMG, people normally say OMG!

Oh My God! Solar energy is so expensive, can't switch to 100% Solar energy as there is not enough space on my rooftop, it requires heavy maintenance, what will happen in the rainy season and finally Solar cell efficiency is not very high and so on.

On the other hand, those who understand the context, concept, and the need for adopting Energy Swaraj fully, will adopt AMG without any qualms.

The problem, however, is that the climate does not have any boundaries. The activities of my neighbour affect me and mine affects him. The activities of one nation affect other nations. Hence the adoption of Energy Swaraj is required for all, across the world, irrespective of whether the nation is developed, developing or underdeveloped if the nation is small, big or very big if the nation is electrified or not electrified and if the nation is rich or poor.

Every individual, household, institution and nation have to take part in adopting Energy Swaraj so that together we can deal with the threat of climate change effectively, efficiently and immediately.

TECHNOLOGICAL VIABILITY OF ENERGY SWARAJ

Are renewable energy or Solar energy technologies available today sufficient to fulfil our 24x7 energy needs, for everyone in the world?

The answer is 'Yes'.

Using the existing Solar energy technologies, if we convert a very small fraction, much less than 1% of total sunlight falling on the Earth, it would be enough to fulfil the energy needs of everyone on the planet.

After applying Avoid and Minimize part of AMG principle, your need for energy will be quite reasonable, even if it is for households, institutions or factories. More often than not such reasonable needs can easily be met locally by using locally available resources like biomass, wind energy or Solar energy.

With the current development in Solar energy technologies, Solar cell technologies and Solar thermal technologies, Solar energy solutions appear to be the best, especially when the need for energy is not very high.

Mostly the Solar cells are made out of Silicon which is abundantly available in the Earth's crust. Among all renewable energy technologies, wind energy technologies and Solar cell technologies have evolved greatly, both in terms of increased performance and decreased cost.

Solar cell technologies particularly are really versatile in fulfilling energy needs across spectrum right from a small Solar watch to big industrial plant. With Solar cell technologies of today, one can power small calculator, Solar lamp, a house, an institution, a factory and even a city. There are cities in the world, which are close to being run on 100% renewable energy like Burlington in the USA, Brasilia in Brazil and Inje in South Korea.

The annual Solar module production has been growing phenomenally with over 30% rate since last 15 years. The efficiencies of the Solar panels are ever-increasing and prices are falling significantly. The associated components of Solar systems, like the battery, power electronics have also been growing in performance and reducing in cost, which is making the entire Solar energy solutions affordable.

THE ECONOMIC VIABILITY OF ENERGY SWARAJ

Are Solar energy solutions economically viable for people? Is the cost of adapting Energy Swaraj high?

The answer is definitely 'NO', especially, if you have followed the AMG principle. If you are a disciplined user of energy and your needs are low, it automatically becomes economically viable in most cases.

"How much would it cost to install a Solar system in my house?" People usually ask me.

It depends on how much you need to generate.

But instead of asking the cost of the system, one needs to ask, what would be the per-unit cost of Solar electricity?

In general, the life cycle cost of Solar electricity, when you are exchanging electricity with the grid without storing in your house, varies in the range of Rs. 3.5 to 7 (5 to 10 USD cents) per unit. And, the life cycle cost of Solar electricity with battery storage comes to Rs. 7 to 10 (10 to 15 USD cents) per unit. While calculating the life cycle cost, the replacement and maintenance cost of components is accounted over the life-cycle of the Solar system, which is usually taken as 20 to 25 years.

If you compare these costs of Solar electricity, with the cost of domestic and industrial electricity in your area, anywhere in the world, more often than not, you will find that the cost of Solar electricity is comparable if not lower than the grid electricity, until and unless the grid electricity is heavily subsidized by the government. Thus, even with the current cost of Solar system components, Solar energy solutions are economically viable. However, it is the high initial capital cost that becomes a financial bottleneck for many.

Establishing Energy Swaraj will not only bring economic benefits to households and communities but also to the nations. Many countries today rely on the heavy import of fossil fuels, as a result, a significant amount of foreign exchange moves out of the country. When people and countries adopt Energy Swaraj and start generating their own energy, they become independent and more empowered, thus economically stronger.

ENVIRONMENTAL VIABILITY OF ENERGY SWARAJ

Would the use of Solar energy results in reduced pollution? The answer is 'Yes'.

Would the use of Solar energy create more garbage in the world? The answer is 'No'.

Renewable energy solution in general or Solar energy solutions, in particular, produce much less carbon as compared to the energy produced by them. As a result, these are referred to as clean energy or green energy solutions. The amount of energy that goes into making of Solar panels, is generated by the Solar panels within 12 to 18 months, while the lifetime of the panels is in the range of 20 to 25 years. The materials used in these technologies are also recyclable. Scientists around the world have been conscious about it and when the volumes are high enough for recycling, it will be done. I am confident about the development of these recycling efforts.

SOCIAL VIABILITY OF ENERGY SWARAJ

All those people who have installed Solar energy solutions in their house or institutions take pride in doing so. Even countries talk highly about the installation of Solar power plants and wind power plants in their country. Therefore, renewable energy solutions are definitely socially acceptable. As we go in the deeper mess of climate change, not only the acceptance of clean energy solutions will increase but also associate pride with it. In fact, it would be great empowerment for individuals and societies, when they adopt Energy Swaraj.

INSTITUTIONAL VIABILITY OF ENERGY SWARAJ

How Solar solutions would be provided to every household in villages, in towns, in every country? How Energy Swaraj can be implemented in a short period?

The whole idea of Energy Swaraj is about being locally self-sufficient therefore the people who provide the energy solutions and those who maintain these clean energy solutions should be locals only.

In my experiments with Solar technology with rural communities, I can see that establishing Energy Swaraj with the involvement of local communities would be the best way, and quite doable too!

BENEFITS OF ENERGY SWARAJ

Adoption of Energy Swaraj, the localized energy generation and consumption can be a great enabler of social and economic development of communities across the world. But it needs to be done in a sustainable manner and should be in sync with the environment. I see Energy Swaraj as one of the most powerful concepts today, having the potential to solve many problems that exist in the world. Energy Swaraj as a tool of production by masses can also be a great equalizer of communities, helping in better resource sharing and can bridge the gap between rich and poor.

The benefits of Energy Swaraj would include:

* **Employment creation:** In Energy Swaraj, the localized energy generation would require more manpower from the local area, as a result, it will create jobs for people in thousands, at locations wherever they are.

* **Arrest migration:** Energy access enables social and economic growth; it can create opportunities for communities to be more productive in their regions. People need not migrate to urban areas in search of better opportunities. In fact, it can result in reverse migration.

* **Skilling manpower:** As the handling of Solar energy solutions, right from assembly to installation to maintenance would require some technical skills, which can be easily imparted.

* **Empowering women:** Not only women but also children by the creation of work opportunities and assured energy supply in the local area, who are important stakeholders in building a sustainable future for nations.

* **Confidence building:** When local people would start assembly, installation, production and maintenance of Solar solutions on their own, it will result in confidence in the technology and technological solutions, thus enabling better, long-term and more efficient use of technology.

* **Energy independence:** Energy Swaraj will make people independent from external energy supply, this independence will also bring economic benefits to households.

* **Economic independence:** With energy independence comes economic independence as a country's currency will be saved from energy imports, which is a huge saving.

* **Ensures security:** Energy is a contentious issue and, sometimes results in tension between countries. This

scenario also puts the country's security and sovereignty at risk. Energy Swaraj can make countries more energy secure.

* **Asset creation:** The enormous asset creation in the local economy will happen as the assembly, installation, maintenance and manufacturing starts happening in the local area which is an essential part of Energy Swaraj. This would catalyse the creation of localized production by the masses in other sectors as well.

* **Brings discipline:** Energy Swaraj brings discipline in the usage of energy, as a result, there will be discipline in the usage of rivers, soil, forests, etc. which would result in protection of all these elements of the Earth's ecology.

* **Mitigates Climate Change:** For the establishment of Energy Swaraj, we will have to use renewable energy sources only, which are more environment - friendly, and helps in mitigating climate change in a major way.

With all the above benefits, who would not like to adopt Energy Swaraj? The time has come to adopt Energy Swaraj in every part of the world.

MISCONCEPTIONS ABOUT SOLAR SOLUTIONS

There are many benefits of Energy Swaraj, though the technologies are available and it is viable to establish Energy Swaraj, I found people having many misconceptions about the Solar energy solutions, when I travelled around the world and spoke to people.

Let me clear out these misconceptions here. However, I realize that clearing the misconceptions and promotion of

Energy Swaraj requires massive awareness campaign in every part of the world, developed, developing and under-developed countries alike.

MISCONCEPTIONS:

* **Low efficiency of Solar cells**

  As a user of technology, we do not bother about efficiency, so why bother about the efficiency of Solar solutions? We need to consider the cost of Solar electricity on a per-unit basis. The efficiency of coal electricity is about 8-10%, whereas the efficiency of commercial Solar module available is about 15-18%.

* **Requires enormous space**

  With current Solar PV technology, one can generate about 100 units of electricity per month from just 100 square feet rooftop area. If you are consuming 300 to 400 units per month, (which is quite high) you only need about 300 to 400 square feet rooftop area. Rooftop space is a problem in the multi-storied building, but then you can generate electricity at some other location, and transmit it to the required location.

  A simple calculation proves that for every country, a small portion of the space i.e. less than 1% would be good enough to fulfil ALL the energy needs of the country.

* **Solar energy solutions, expensive**

  The life cycle cost of per unit of Solar electricity, when you exchange it with the grid without storing in your house, varies in the range of INR 3.5 to 7 (5 to 10 US dollar cents). And, the life cycle cost of per unit of Solar electricity when you are storing energy in battery comes to INR 7 to 10 (10 to 15 USD cents). While the actual cost that people pay both for domestic purposes and in the industry are higher than the costs of Solar electricity at per unit basis in many parts of the world.

In fact, there are many business people who are willing to install Solar solutions on the rooftop of institutions at a zero investment, the institutions can buy Solar electricity at a lower rate than the current rate of electricity paid by the institutions. In a way, it is a win-win situation for both.

* **Require significant maintenance**

In case of Solar technologies, the only maintenance required is to clean the dust that settles on the Solar panels. As per definition, PV modules degrade only by 20% in its lifetime of 25 years. After its lifetime, the Solar panels would still perform up to 80% of its initial capacity.

In terms of electricity storage, with every Solar PV system, you do not need battery storage. Under net metering, one can connect the PV system to the electricity grid and the grid becomes battery storage. For standalone PV system, one needs batteries, and even maintenance-free batteries are available. These days batteries with a lifetime of 5 to 7 years are easily available.

* **What happens during the rainy season?**

The Solar panel output depends on the input. As long as some light is falling on the panel, there will be some output. In most parts of the country, except coastal areas, 50-80% radiation falls during the rainy season as compared to sunny days, hence Solar panel's output would be 50-80%. On full overcast days, it would be in the range of 30-40%.

I strongly believe that Energy Swaraj is feasible, possible and achievable in every part of the world.

PART - G

LET'S CREATE A PUBLIC MOVEMENT –
GOING GLOBAL

In the times of Mahatma Gandhi, the Charkha became a symbol of change and in the times of climate change threat, this Solar Lamp (SoUL) will become a symbol of change.

– Dr. Anil Kakodkar, Scientist & Former Chairman, Atomic Energy Commission

CHAPTER 20

MY GANDHI GLOBAL SOLAR YATRA (GGSY)

The only constant in life is the change.

– Heraclitus

Nothing is permanent here. Everything around us is changing; people are changing, seas are changing, trees are changing, surroundings are changing, thoughts are changing and above all the climate is changing, everything is changing, every moment.

Even I am changing…

…BUT I will change so much? I could have never imagined even in my dreams, that I, Chetan, a professor from IIT Bombay, would someday undertake Gandhi Global Solar Yatra (or journey). This was an unimaginable idea by anyone, including myself, until I embarked on it on December 25, 2018, from Sabarmati Ashram, Ahmedabad. Probably, Gandhiji's Dandi Yatra inspired me, which proved a milestone in India's freedom struggle. Maybe my Gandhi Global Solar Yatra also proves a milestone for people of this world, who may attain energy independence, Energy Swaraj, who knows?

Around 20 years back I started my journey in the field of Solar by travelling to Europe for my doctorate studies, a journey which took me beyond boundaries of not only my nation but of my thinking, exposure, vision and even my consumption of resources. Now, 20 years after that outward journey, I took another journey, the Gandhi Global Solar Yatra, a yatra, that helped me in expanding my thinking, my exposure, my vision and even further, driven me to reduce my consumption of resources. This yatra has expanded my vision, both inwards and outwards.

THE IDEA OF THE GLOBAL SOLAR YATRA

I remember it was 8th October 2018, and I was sitting in my SoULS project office, after organising a very successful event on 2nd October - Gandhiji's Birth Anniversary day. The event was called Students Solar Ambassador Workshop, wherein we trained about 1,32,000 students to make their own Solar lamps, covering almost every state of India. It turned out to be a very successful event, which also registered itself in the Guinness Book of World Records.

While we were still basking in the success of the workshop, on 8th October 2018, the Intergovernmental Panel on Climate Change released its report on climate change. I was struck by its recommendations, the actions that would be needed to limit global warming to 1.5-degree centigrade. The report recommended a rapid and far-reaching transition in energy. Even bigger, it recommended the world to become Net-Carbon Zero by 2050.

I questioned myself, *"How on Earth, one can achieve this?"*

And this question has stayed with me for many days.

I thought, *"I need to do something about it."*

Our ongoing Solar Urja (energy) through Localization for Sustainability (SoULS) project had reached several million households, but I felt that this is not enough. I need to do something more, something bigger.

After the celebration of 149th birth anniversary of Mahatma Gandhi on 2nd October 2018, Government of India announced the plans for celebration of 150th birth anniversary on 2nd October 2019, a very important occasion, not only for the people of India, but also for the world, as 2nd October is also celebrated as International Non-Violence day by the United Nations.

Now, I could view many things coming together like our successful Solar project involving local communities, successful Student Solar Ambassador workshop, disturbing IPCC report on climate change and Gandhiji's 150th birth anniversary celebration.

Wow! I suddenly jumped out of my chair!!

Something struck me, it filled me with excitement.

"Why shouldn't I take up a Global Solar Yatra to promote Solar energy to help mitigating climate change?" I questioned myself.

It made enormous sense to me.

I thought *"What could be a better tribute to Gandhiji on his 150th birth anniversary than a Global Solar Yatra when I go around the world to promote Solar energy through the involvement of local communities and helping climate change mitigation."*

We know that Mahatma Gandhi and other leaders have taken up such yatras for public causes.

I discussed this with Swati, one of the key persons for ensuring the successful organization of Student Solar Ambassador of 2nd October 2018. She also thought it was a great idea. I discussed it with some more people, it sounded like a great idea to everyone!

After quite a few deliberations, the name of the yatra got fixed: Gandhi Global Solar Yatra. It captured all essential messages that I wanted to give; Gandhian philosophy for a sustainable world, Global outreach as the climate has no boundaries, Solar energy as Sustainable energy solution and reaching out to all by travelling to them.

The target was fixed for visiting at least over 25 countries and appoint over 1 Million Student Solar Ambassadors across the globe by giving them hands-on training on 2nd October 2019, the great occasion of 150th birth anniversary of Mahatma Gandhi. Since the very beginning I knew, time was too short to achieve this phenomenal scale, but then I thought Gandhiji's 150th birth anniversary will come once in my lifetime, so it was *now or never* situation and I decided in favour of *'now'*.

Then came a big question; *Who will fund this yatra? And why?*

Somehow, I thought that the government should not fund it, as the purpose of the yatra was to create awareness amongst the people of the world. Therefore, I thought that people should fund it, individuals as well as Corporates through their CSR funds. Somehow, I have learned that for any good work funds eventually get available, though there was no clear path or idea where these funds will come from, I was sure that the funds will come.

Therefore, without really worrying about the arrangement of funds that were required, I started work on my yatra.

The name 'Gandhi' was in my yatra, and unknowingly, I realised that Gandhi became part of my thoughts and actions. I started to read him more...

...AND, THE YATRA COMMENCED!

What could have been a better place than Sabarmati Ashram in Ahmedabad to start my yatra where Gandhiji lived for several years? From here only, Gandhiji started Dandi march, known as Salt *Satyagrah* which had a significant impact on the Indian Independence Movement.

On the evening of 25th December 2018, I reached Sabarmati Ashram, Ahmedabad, where I had a wonderful opportunity to stay in the guest house of the ashram.

25th December was Christmas, indeed a wonderful day to begin my Gandhi Global Solar Yatra, and I prayed Lord Jesus to bless the moment. Everything here was appearing to be very close and connected, all my thoughts and actions had gathered around a purpose - *How to make Gandhi Global Solar Yatra, a people's movement?*

In the evening, while I was sitting in the *verandah* of the ashram where Gandhiji and Kasturba lived, I closed my eyes to imagine and recreate those times when Gandhiji was spearheading Freedom movement, joined by the greats like Vinoba Bhave. The scenario would have been really charged towards the movement which would have involved great leaders, freedom fighters and the public at large for India's freedom. The feelings crossing my heart, the feeling of involvement, the feeling of a connection, can't be expressed in words...

...I felt one with the movement started by Gandhiji some 100 years ago. This time the movement is about protecting the environment and promoting sustainability of life on the planet using a non-violent form of energy, the Solar energy.

Non-violence was an integral part of Gandhian movements.

After an exciting evening, it set me thinking: *How can we equate the usage of Solar energy to non-violence to the environment?*

ENERGY SWARAJ COINED!

My thinking, firmed up further, as I was reading Gandhiji's book on Gram Swaraj. While thinking overnight, it struck me that, if the energy is generated and consumed locally, if individuals and society generate their own energy using Solar energy, then, can it be termed as *Energy Swaraj*?

Energy Swaraj would take the lead from the path shown by Gandhiji's Gram Swaraj, which propagated that all resources used by the community to be generated and consumed locally like food, clothes, etc. Similarly, can we have Energy Swaraj where all energy needs of the community are generated and fulfilled locally?

The concept of non-violence is universal and is applicable everywhere. What if we apply the concept of non-violence to the environment, owing to the high degree of emissions leading to pollution of alarming proportions? Throwing dirty and toxic materials in rivers and seas, thus not leaving the planet liveable for future generations. I think this is violence towards the environment.

The human act of causing water, air and environmental pollution, actually is nothing short of utter violence to nature, which is quite rampant now. I refer here a discussion with Dr. Janak Palta, a Padma Shri Social Worker, who termed this violence as the rape of nature. I was stunned to hear these words, but on careful analysis, I found this to be true in some sense.

During my Gandhi Global Solar Yatra (GGSY), if I am able to establish this fact that our actions are great violence to nature, and if it gets accepted by the world at large, the yatra would achieve phenomenal dimensions.

TIMES OF INDIA BESTOWS THE TITLE "SOLAR MAN OF INDIA"

Ms Yogita Rao from The Times of India did my telephonic interview on the eve of my yatra and asked me, *"What is the specific purpose of your global yatra?"*

I emphasized the fact that though there is awareness in the world about climate change, there is not enough action in the direction to arrest that.

I told her, *"The promotion of localized energy generation and consumption was the main purpose of my yatra, but it was not the only purpose. I wanted to begin some kind of movement with my yatra, not only for spreading the awareness but bringing an "actionable awareness" to make this a successful movement - Energy Swaraj movement."*

I explained further, *"Bringing millions of young minds on-board all across the world for self-energy generation and consumption, by giving them hands-on training to make their own Solar lamp was also an important objective of my yatra."*

A few days later, the article appeared in The Times of India with the title "Solar Man Solanki starts a mission across continents." Many have been referring me with this title in the past, but getting referred like this from the most respected and circulated newspaper of India, was indeed a great honour.

THE UNFORGETTABLE MOMENT OF MY LIFE

On the night of 25th December, I slept thinking that I will wake up early in the morning to be in the Ashram for the morning activities and be in sync with Gandhiji's principle of starting the day with *Prarthana* at 4 am. Though Ashram opens only at 6:30 am, still, I did not want to miss the opportunity to be in the Ashram to feel how Gandhiji spent his mornings.

With these thoughts, I reached Ashram and headed straight to the place where Gandhiji and Kasturba lived. I started my *sadhana;* followed by *Surya Namaskar, Pranayam*, Meditation and Jogging.

While in meditation, a thought crossed my mind, *"I should do some Seva (service) in the Ashram."*

Then I saw caretaker getting ready to clean the building. By now the daylight had already broken.

"I want to do some service here, please give me some work," I asked the caretaker.

He took a pause for about 10 seconds, and replied, *"Why don't you clean the Gandhiji's room?"*

The public access is not there for Gandhiji's room, people could see it from the outside only, through the window, and here I was getting an opportunity to clean the room. Wow! My happiness knew no bounds, my wish was granted. Ecstatic, that I was, I opened Gandhiji's room, and cleaned and mopped the room - the personal sitting space of Gandhiji and his *Charkha* too!

It was truly a fulfilling moment as I felt Gandhiji himself was blessing me. The time spent there was just awesome, the touch of the walls and floor was amazing...

...Unforgettable! I felt privileged.

What a beginning of my Yatra!

THE YATRA – A LIFE-CHANGING EXPERIENCE!

What an amazing, life-changing yatra it turned out to be, the Gandhi Global Solar Yatra. The yatra of the world, with a purpose to promote Solar energy, the Energy Swaraj and encourage communities to enroll young students from schools and colleges to take part in the planned global event of 2nd October 2019, the Student Solar Ambassador workshop.

I have been capturing every single day of my yatra and writing experiences in the form of blogs. It would be difficult to write everything about the yatra, but I can say that it was an eye-opening, satisfying and vision broadening experience. It strengthened the SOLAR truth and left me with some invaluable lessons for life.

I will enumerate a few pointers, which are most relevant for the book's readers.

While I was travelling across the world, in different time zones, the team in my office made all the arrangements like fixing meetings, accommodation, pick-up, drop, whom to meet, when to start from one place, when to reach to next place, everything. I must acknowledge that the team back in Mumbai was doing an excellent job.

How lucky I am to have such a dedicated team led by Swati Kalwar and her team members Nikita Arora Linihar and Harshad Supal along with Vinit, Ameya and Meghen.

As every single day of the yatra was precious, I attempted to make the most out of it. The plan was to have four meetings or seminars every single day, hence just about two hours for each meeting or seminar. This was to be done in a new place, with new people, with a new language, new climate, new customs, new food, and new transportation system, and to be done across the world, every single day. However due to logistics and other reasons, sometimes, it was not possible to have four meetings in a day.

Many times, the day's work would include travel from one city to another, one country to another. It was tough, very tough, many times it was very taxing on the body, as well as on the mind. But I survived, I did not fall sick or missed any meeting.

Here, my background as a practitioner of Yoga and Meditation was useful. Being a teacher of the 'Happiness Program' of Art of Living was energizing and at several places, I met people belonging to Art of Living. It was like finding family members in every part of the world.

What an amazing work Gurudev Sri Sri Ravishankar has done, uniting the world. Other than Mahatma Gandhi, if there is anyone person whom I remembered most during my yatra, he was Gurudev Sri Sri, for thanking him in good times and remembering him in tough times.

I travelled to many countries in a short span of time, in many places I would stay for just one or two days. It meant that things were changing very rapidly, right in front of my eyes, as if I was watching a movie, with scenes on the screen changing one after another. In one day, I would be somewhere in the middle of snow and some other day in the desert, someplace will be cloudy and raining and some other place dry and sunny. One day people around me would have dark skin, on another day, white and next day somewhat in between. It all changed so quickly. As time was a constraint, this was the only way to cover the world.

I travelled to the US covering California, New York, Boston, Washington, Europe: Germany, France-Paris, Belgium, Brussels, Berlin, Switzerland, Spain, Portugal, Africa: Mauritius, Tanzania, Nairobi, Ethiopia, Benin, Uganda, Rwanda, UAE: Dubai, Oman, Doha, Abu Dhabi, South East Asia: Bali, Jakarta, Indonesia, Thailand, Malaysia and Myanmar, South Asia: Bangladesh, Nepal, Sri Lanka, and far-flung Brazil in South America.

In India, I have travelled to New Delhi, Bangalore, Chennai, Ahmedabad, Surat, Vadodara, Bhopal, Indore, Jaipur, Udaipur, Dungarpur, Andaman & Nicobar Islands, Silchar, Agra,

Aligarh, Kanpur, Lucknow, Meerut, Varanasi, Mandi, Dalhousie, Dehradun, Roorkee, Thrissur, Cochin and many more!

During my yatra, I have travelled about 150,000 km.

I used to carry along Solar lamp kits, particularly during my visits abroad where I conducted training programs also.

CONCLUSION OF MY YATRA

As all good things come to an end, so did my Yatra. I started my yatra from Sabarmati Ashram on 26th December 2018 and concluded at the same place on 15th August 2019, on Independence Day. Gandhiji's stay at Sabarmati Ashram and its activities played an important role in India's independence. 15th August was doubly auspicious, as it was also the day of the festival of Raksha Bandhan in India.

Not only that I ended up my yatra on 15th August but we also officially inaugurated Energy Swaraj Foundation in the presence of Dr. Anil Kakodkar, that too from Sabarmati Ashram. The thought of establishing the Energy Swaraj Foundation was there for a long time. Finally, my friend Venkat Rajaraman and I co-founded it to begin the action.

After completing official formalities, on 15th August we got it inaugurated. The purpose of the Energy Swaraj Foundation is to act as a platform for bringing Energy Swaraj, across the world. Hopefully, millions will be part of the Energy Swaraj Foundation. Hopefully, this foundation will lead the Energy Swaraj Movement across the world.

Like at the beginning of my yatra, another great coincidence happened, when I was concluding my yatra at Sabarmati Ashram. While I was giving media interviews in the premises where Gandhiji used to live, someone came and said, *"I will open the room of Gandhiji for you, you can go inside and take pictures with original Charkha of Gandhiji."*

This way my yatra started as well as concluded being in Gandhiji's room at Sabarmati Ashram. Both times, it happened without me asking for it. In this way, I was lucky to enter Gandhiji's room before commencing the yatra and after finishing the yatra.

I felt again that this is a blessing directly from him.

KEY LEARNING POINTS OF YATRA

Travelling around the world as a part of Gandhi Global Solar Yatra was an incredible experience. Seeing the world in one go, having the opportunity to look at the entire spectrum of the world, developed countries, developing countries, and under-developed countries in the same time frame was quite an experience. The yatra had a deep impact on me and based on the experiences, some invaluable conclusions could be drawn which will have a long-lasting impact on my thinking, and on the planning and execution of Energy Swaraj movement across the globe, some of these conclusions may still be evolving.

TEN MOST IMPORTANT POINTS ARE ENUMERATED BELOW:

1. Mitigation of climate change is on nobody's mind. It would not be wrong to say that more than 99.9% of people across the world are not sensitive towards this issue. There is a great ignorance of this call that 'some serious action is required' everywhere in the world.

2. There is an acceptance of the concept of Energy Swaraj, as most of the time after my talk, the host would tell me, if

there was a time, we would have organized it in a much bigger way. However, from the follow-up questions and discussions, I realized that it is hard to absorb Energy Swaraj's idea in one lecture. It is hard for many people to imagine that all their energy needs can be generated and fulfilled locally.

3. There are many misconceptions in the world about Solar energy, hence a lot of awareness is required towards Solar energy.

4. I was surprised to note that the off-grid Solar solutions or standalone Solar solutions have spread less than the centralized Solar power plants, probably a sign of limited, short-term vision of people, world-over.

5. Not only common man but even leaders don't think long-term, even if they do, they turn a blind eye and fail to understand the seriousness of climate change.

6. The need for Energy Swaraj is very obvious and immediate, I realized this during my travels across the length and breadth of the world. I became comfortable to see less development in places, which meant less wastage, less damage to the environment, thus living more in sync with nature.

7. Traffic jam is a reality in all major cities across the world in rich as well as poor countries. I would think, "how come every city has failed to plan properly?" I realized that it is not the problem of planning, it is part of some major flaws in our thinking and living. Even the current economic model of the world is flawed, which I believe is also responsible for climate change.

8. The world is a crazy place, I could see the gap between rich and poor everywhere in the world, in Gandhian terms, it is a kind of violence.

9. Every human being on the planet is the same but with different face masks. People are driven by similar desires, aspirations and behave almost in the same manner in a similar situation. People appear to be different, but their core values are the same everywhere in the world.

10. The world is a bigger version of India, the land of one-sixth of humanity has everything that the world has to offer, riches, deprivation, ultra-modern, basic, classy, massy, simplicity, complexity, culture, religion, metros, villages, etc. whatever is there in the world, is there in India.

I HAVE CHANGED

It is said that as a human, it is difficult to change. We keep behaving in the same manner, but I feel that I have changed. I have certainly become a different person after my Gandhi Global Solar Yatra, in terms of my thinking, at least, if not action!

I had a different world view before I started my yatra, and now it is different. I see that less development is all right or even good, as it means that people are living in sync with nature without inflicting violence on it. However, I am fully aware that people with fewer opportunities, would immediately jump to the other side if given an opportunity.

The glamour of developed nations doesn't impress me anymore, in fact, it looks farce now. I see that the world has gone in one direction only, the progress of humans is measured in terms of economic progress only, the other aspects like humanity, spirituality and morality have all taken a back seat.

Whatever Mahatma Gandhi preached the world, he adapted those things in his own life, he practised them first. I have read and spoken so much about Gandhiji while travelling, that his ideals automatically come in my thoughts and many times in actions as well.

After coming back to India, at home, we have surrendered our refrigerator, washing machine, microwave and geyser. We never installed AC at home. Probably for my wife Rajni, surrendering the refrigerator was difficult, but eventually, my children and wife became at ease with it. This is simply adopting 'A' of the AMG principle.

As the days are passing by, and I am speaking more and more about Energy Swaraj, explaining how climate change is affecting the lives of people, my resolve to promote Energy Swaraj is becoming stronger by the day, and my commitment to work for this cause is becoming stronger.

This yatra has brought me to this juncture, let's see where the future yatras take me. Let's see whether dark energies prevail or bright energy of the Sun takes the centre stage.

After all, the world is nothing but a play of good and bad energies!

CHAPTER 21

SOLAR ENERGY AS A UNIFYING, EQUALIZING FORCE: RECOMMENDATION FOR COUNTRIES

One of the things that troubled me a lot while travelling across the world as part of my Gandhi Global Solar Yatra is noticeable, unforgettable and very unfortunate 'inequity' that exist everywhere in the world: within countries, within continents and between continents.

I could see an unequal world within the New York City Subway in America, while walking through the streets of Adis Ababa in Ethiopia, while driving on the roads through Rio de Janeiro in Brazil, between the train stations of Brussels in Belgium and on the roads of Banda Aceh of Indonesia. I know very well, how an unequal world exists in the island city of Mumbai, India.

This inequity in the world is the product of the economic model that we have adopted since industrialization. Ever since inequity in the world is on the rise. It is also clear that this inequity within a country and between the continents is the cause of nearly all major problems, the world is facing today like violence, poverty, education, exploitation of people, misuse of natural resources, and of-course climate change.

There are absolutely no ways where rich people and poor people, rich countries and poor countries can have a peaceful coexistence. I strongly believe that the solution to most of the problems the world is facing today lies in creating equality and increasing the belongingness amidst people, as it leads to an attitude of caring and sharing.

We need to follow the philosophy elaborated copiously in Indian scriptures of वसुधैवकुटुम्बकम् or Vasudhaiva Kutumbakam, which means that *"the world is one family"*.

For the Sun, the world is one family. It shines everywhere on the planet and nurtures everyone equally. One thing which is common to every creature on the planet is the Sun, the source of energy for all. Irrespective of the location and country, everyone receives Solar energy without prejudice. I believe that not only Solar energy is the great equalizer but also a great unifying force in the world.

If we all make, Solar energy a base of our lives, a driver of our social and economic growth, not only will the world be a more equal and peaceful place but also, it will help in mitigating the major existential threat to humanity looming large due to climate change.

Fortunately, technological progress and economic viability of renewable energy solutions in general and Solar energy solutions, in particular, allows us to base our lives mostly on renewable energy, if not completely.

Based on the personal experiences gathered during my visits to several countries, based on my understanding of the various countries and of the technological advancements in the Solar energy sector, I have certain recommendations for countries based on their current status of electrification.

In general, we all need to understand that the current economic model, which propels the world continuously towards a higher level of consumption of everything like water, food, energy, material, etc. which in turn requires a higher level of resources, and a higher level of energy consumption, is not sustainable.

Ever-increasing consumption can NOT be a model for sustainable life!

Let's take a pause, put a thought to it, apply arguments to it and accept it. There is no way the Earth can support the ever-increasing level of consumption; it's writing on the wall!

Recommendations for countries world over:

100% Electrified nations:

Countries like the USA, Germany, France, Belgium, Oman, Malaysia, UAE, etc. provide nearly 100% electrification access to its citizens who enjoy all modern amenities. There are well over 60% of the countries of the world which fall in this category.

Mahatma Gandhi wrote in Young India on 30th April 1931 that *"I dare to announce that people from Europe are becoming slaves of comforts of life, and if they do not want to bury themselves under the burden of it, European people need to change their outlook (towards the modern life)."*

Contrary to the aspects of modernity and comfort, he believed in *"simple living and high thinking"*. I think one and all need to adapt to this philosophy as he also said, *"There is enough on the Earth for everyone's need, but not for everyone's greed."*

Countries with nearly 100% electricity access to its citizens must cut down significantly on the resources they use. I have seen during my yatra to such countries, a very high level of energy consumption, rather huge wastage, which are an integral part of life.

Someone in the Middle-East told me very proudly that we have an efficient system of collection, segregation and processing of garbage. In response to this, I asked him, *"Why do you create so much garbage in the first place?"* He had no answer to this.

Technology can not and will not be the solution for each and everything that we are doing. We really need to understand this and go back to the basics of sustainability, that is practising discipline in consumption. We should make it the first principle of sustainability.

In the wake of climate change, wasteful use of fossil energy can only be considered a near-criminal act. Mere sensitivity towards the issue in hand can help the countries to cut down on energy needs significantly.

If the people in these countries start following AMG (Avoid-Minimize-Generate) principle, it will be a great gesture and respect towards nature. With very good energy infrastructure, these countries, then need to integrate renewable energy solutions and move towards 100% renewable energy-based economy, on the lines of Germany and Sweden, but much faster than them.

With greater public participation, the way every one generates and fulfils their own financial needs, here everyone can take responsibility to generate and fulfil ALL their energy needs (including electricity, transportation, cooking, etc.).

I believe the transition to 100% renewable energy in these countries can happen and need to happen in the next 10 years, let's say by 2030. In my own sense, most of the citizens in these countries would be able to achieve the goal, without financially burdening the government.

Nations approaching 100% electrification:

All those countries which are approaching near 100% electrification like India, Serbia, Nepal, etc. need to take a pause on their current approach of electrification. As the world needs to take a U-turn from the centralized fossil fuel-based energy generation and consumption, and since the remaining target is small, these countries, can show by experiments, the effectiveness of the use of Solar energy, by adopting 100% Solar energy to provide electricity to the remaining people.

The remaining population, in these countries, most likely will be living in either difficult geography or having low population density or having low-income activities or a combination of all these factors. These governments can save enormous cost by not following the centralized model here. Their efforts on 100% adoption of Solar energy, with the involvement of local communities, can pave the way for the rest of the world.

Of course, the other population (having electricity access) to adopt AMG principle, as quickly as possible, and should start the reverse process. This will make their economies stable as it will lead to job creation at the local level and in turn strengthening of the local economy.

Nations with 60-80% electrification:

Then, there are countries like Kenya, Myanmar, Senegal, Libya, Ghana, Pakistan, etc. where electricity access to its citizen is in the range of 60 to 80%. Here providing access to energy to all its citizens should be the first priority as access to energy is a great enabler of growth in life.

Staring with AMG principle, the governments can start building infrastructure that promotes localization. With little efforts from the people, government and external support, the change can be driven in a manner that energy generation and consumption move towards the localized model of attaining Energy Swaraj. Creation of skills in the management of Solar energy technology solutions can greatly enhance the overall skill set, which would be useful in other trades as well. Creation of energy assets in the local economy would be useful in the promotion of other technological solutions and asset building in the local economy.

Nations with less than 50% electrification:

Countries like Uganda, Tanzania, Ethiopia, Rwanda, Madagascar, Namibia, etc. where electricity access to its citizens is around 50% or even lower, providing access to energy to all its citizens should be the first priority. I also know that in these countries, many people struggle for two meals a day. But access to energy, or let's say, access to some energy can start immediately, without adding significant cost.

Our SoULS program has demonstrated how with little investment one can involve local people to assemble Solar products for their own needs, creating confidence in technology and jobs for local people. This trend of localized energy generation and consumption can help the countries greatly to upgrade themselves and become self-sufficient. I believe that AMG principle would automatically be built in these countries, in their efforts to generate and consume energy at the lowest possible cost.

These countries can develop a completely decentralized model for electricity generation and consumption and can become a model for the rest of the world, as they have the opportunity to establish the most efficient economies. They would need financial support in providing energy access to all.

The rest of the world should be happy to provide support, as it will help in the mitigation of climate change and making the world a better place.

From my experience and experiments, I am confident that the recommendations presented here can provide countries and its people, not only sustainable energy access without harming the environment but also it will provide several other benefits like job creation, women empowerment, strengthening of the local economy by way of asset creation, making people and country energy independent and above all mitigating climate change.

This way we see that Solar energy can be a great unifying and equalizing force in the world.

CHAPTER 22

GLOBAL STUDENTS SOLAR AMBASSADORS WORKSHOP: THE BEGINNING OF ENERGY SWARAJ MOVEMENT

Who wouldn't love his children? Who wouldn't plan for their bright future? Which country wouldn't ensure a bright future of their young ones?

The young generation is the future of every nation in the world. It is for them that the parents work hard, it is for them that they try to build a bright future. It is for the bright future of their children; parents do what they are doing. Isn't it?

Greta Thunberg, a 16-year-old Swedish girl, who is asking world leaders to take strong action to tackle climate change, told recently to a world leader; *"You say you love your children above all else and yet you are stealing their future in front of their very eyes."*

Our love for our children is not real love, our love is blind, as it views their future with a short-term perspective. In the current context of climate change, when we look at the immediate future of the young generation, it looks bleak. And, when we look at it in relatively long-term, when there is an estimation that by the end of this century the global temperature may rise to about 4 to 5 degrees above normal, the future of young ones looks scary, such is the threat of climate change.

I keep telling people in my talks that those who are in their 40s and 50s, most likely, will not be there on the planet after 40 to 50 years from now. But the young ones, who are in their teens will be there. They are going to bear the burden of climate change and would have to pay the price of climate change, eventually, they will face the consequences of climate change.

If this is the case, then the question is, *"What kind of love we have for our children, for our future generations?"*

Are we cheating with young generations?

THE YOUNG GENERATION, AN IMPORTANT STAKEHOLDER

I strongly believe that the young generation of today should be made an important stakeholder in the world energy scenario. Since our life is driven by the use of energy, and its use causes climate change, the choice of the model of energy generation and consumption should be given in the hands of youngsters, as it is their future, which will get affected most.

In our Solar Urja Lamp (SoUL) program, the local women get training to assemble and repair Solar lamps. More than 9000 women from rural areas have taken part in Solar lamp assembly program and have earned money from the job. From this example I can say that that making a Solar lamp is not rocket science, anyone can assemble it.

I have been thinking since a couple of years, *"Why not students? Why don't we give training to school and college students to make their own Solar lamps?"*

As a trial, I conducted training in two schools for over 400 students, in Maharashtra and Madhya Pradesh in India. The response was very encouraging in both the schools, the level of interest, energy and enthusiasm showed by the students was really great.

One of the most exciting parts of this training was the evening time when the workshop would conclude. It was the time when all students would gather with the Solar lamps made by them, brimming with high energy, and switch on their lamps together. The moment, when all the students gathered in the hall and switched on their lamps, there was huge joyous roar in the hall! I still remember that powerful moment.

This is the might of doing something on your own. I guess that is why Mahatma Gandhi emphasized on the training of not only Head and Heart but also training of Hands (Three Hs). As I travelled across the world, almost everywhere I saw the lack of skill in doing things by hand. I saw, it was difficult for many students to use even a screwdriver, which was required to assemble the Solar lamp.

After these successful experiments of training a large number of students to make their own Solar lamps, the next idea was *"Why don't we train every student for making their own Solar lamp?"*

Even better, *"Why don't we make every student an Ambassador of Solar energy?"*

It sounded like a great idea to me. A student would become an ambassador of Solar energy, if he or she learns to make their own lamp and most importantly use the lamp for study purpose, irrespective of the status of the student; whether rich or poor, having access to electricity or not, living in America or Africa.

If students, across the world, start making their own lamp and start using it, they will be making a very strong statement to world leaders. I believe, from this action of students, the world leaders will be compelled to take action towards mitigating climate change.

I imagined that the training and use of the Solar lamp by students should be as easy as the use of a mobile phone today. Every child across the world, can play with mobile phones and uses them efficiently, similarly, every child across the world should be able to play with Solar devices and use them efficiently. So, the Solar devices should be in straight competition with the mobile phone! Interesting! Isn't it?

GLOBAL STUDENT SOLAR AMBASSADOR WORKSHOP

I prepared a plan to train 5000 students to make their own Solar lamp on 2nd October 2018 and shared it with my team. The plan was made a few months in advance. Slowly the plan expanded, several other locations where our Solar lamp project was running also joined in for participation for the 2nd October event. We ended up training more than 1,32,000 students all across India with over 850 schools participating. It was a very successful and well-appreciated event. The students enjoyed themselves everywhere.

More than 5700 students gathered in IIT Bombay and learned to make their own Solar lamps. In the evening when all the students gathered on the ground as a part of conclusion ceremony, and when they switched on their lamps, the whole ground filled up with the light as well as with the roar of the students. I witnessed the energy of 5700 students and the power of the Solar lamp in the hands of young ones.

In fact, on the very same evening, there was a function in Delhi, where I was supposed to present the various demonstrations of our projects, to be visited by Prime Minister of India and United Nations General Secretary. I had to make a hard choice of being in Delhi or in Mumbai. I chose to stay back in Mumbai. Witnessing the 5700 students lighting their lamp was worth staying back. The event found a place of pride in the Guinness Book of World Records.

Based on the powerful experience of 2nd October 2018 and considering the fact that 2nd October 2019 will be celebrated as 150th birth anniversary of Mahatma Gandhi, I thought of conducting this Students Solar Ambassador workshop, all over the world. Initially, the target was set to conduct the event in at least 50 counties and train at least 1 million students, which is going to be mammoth operation by all standards.

It required reaching out to 50 countries, training the teachers locally so that they can conduct the event on 2nd October 2019, ensuring that material reaches everywhere in time, that too without any funding support from the Government. Above all, doing it in just 10 months' time. I undertook the Gandhi Global Solar Yatra for this purpose.

While I am writing this book, the preparations for organizing the Global Student Solar Ambassador Workshop are in full swing. The task still looks uphill but I am confident that with a very energetic team working with me, we will be able to pull it through well. As of now, there are more than 100 countries on board and we will definitely be expecting to train much more than 1 million students across the world.

With my Gandhi Global Solar Yatra, the tiny Solar lamp has reached nearly every corner of the world, and with this Global Student Solar Ambassador workshop on 2nd October 2019, the same Solar lamp is going to reach to millions of students all across the world. It will sensitize the young minds on the need for clean and sustainable energy and dangers of climate change. It is a small Solar lamp, but I believe it is a very powerful device, symbolizing change! It firms up the need to adopt Solar energy for climate change mitigation and building a bright future for young ones.

Dr. Anil Kakodkar, well respected Indian Scientist and Former Chairman, Atomic Energy Commission of India, while addressing those 5700 students on the 2nd October 2018 said, *"In the times of Mahatma Gandhi, the Charkha was a symbol of change and in times of climate change threat, this Solar lamp (SoUL) will become a symbol of change."*

It was such a great honour for the Solar lamp to be compared with Gandhiji's *Charkha*. This comparison tells us how powerful a small Solar lamp could be!

This Solar lamp can spark a movement, a movement towards greener, cleaner and sustainable future.

Through this book, I would like to appeal to each and every student across the world to become Student Solar Ambassador. If you do not get the opportunity to do so on this 2nd Oct. 2019, never mind. Find earliest possible opportunity to learn to make your own Solar lamp and start using it for your studies. This will definitely create lot of pressure on the old boys to take right action, for the planet and for your future.

Through this book, I would like to appeal to all parents to provide an opportunity to your children so that they can learn to make their own Solar lamps and other Solar products. I assure that it will be life-changing experience for your children. Such gesture from your side will have power to change the direction of energy generation and consumption for the betterment of the planet and for the better future of your children.

CHAPTER 23

LET'S CREATE ENERGY SWARAJ MOVEMENT

Life on the planet Earth would not be sustainable in the current energy scenario, where excessive usage energy is rampant world over. This is leading to rise in Carbon footprint in alarming proportions, which would make the survival of the human race questionable, going forward.

Do we have an answer for this?

In this book, I have discussed at length how the current energy generation and consumption system of the world would not be sustainable in the long-run. There are many shortcomings in the current energy system, the supply itself is at fault, which is causing climate change all over the planet.

WE NEED A GLOBAL PUBLIC MOVEMENT

Tough times demand tough measures. I believe that the world is going through very tough times, as I am writing this, there is a heatwave in the US and Europe, there is a serious threat of entire Arctic Ice Cap melting by 2030, there are serious threats of floods and droughts, everywhere in the world. The world population is still growing but natural resources that support life are depleting. How the people in the world are going to sustain amidst all this?

From my experience of two decades, from my travels under Gandhi Global Solar Yatra around the world, from my reading about the developments around the world, from the actions that policymakers and governments world over are taking to mitigate climate change, I have come to the conclusion that overall efforts to mitigate climate change are minuscule, while the destruction caused due to climate change is alarming. With a rise in production and consumption everywhere in the world, things are becoming even worse.

I have come to the conclusion that Governments, due to their short tenures (4-5 years) and need to respond to immediate concerns of the people, and can't really afford to think seriously in the long-term.

The Sustainability requires long-term thinking, say for 50 years, 100 years, 500 years which Governments, across the world, cannot afford. Therefore, the right actions that are required to tackle climate change are not being planned properly.

I strongly believe that we, the world, needs the contribution of every individual to take care of this mega problem standing right in front of us. We ALL, each one of us, need to first understand the scale of the problem, then act upon it in the same or on a bigger scale to solve the problem.

If each one of us, start educating ourselves about the amount of resources that we are consuming and the actual amount of resources that we should ideally be consuming for a sustainable world, we can address the problem. If we understand that our use of energy is a major culprit of climate change and start appreciating the invaluable and irreplaceable habitat we live in, we will be in sync with nature. With such awareness and action, we would be able to mitigate climate change.

Not only we would be able to tackle this life-threatening problem, but also ensure a bright future for our children and grandchildren. This requires a contribution from each of us, every citizen of the world, from every corner of the world, ALL of us, together ONLY, we can address this problem.

We need a public movement on a global scale.

Let us stop cutting the very same branch on which we are sitting, let us stop destroying the very same environment in which we are living, let us stop ill-treating the planet which has limited resources, let us stop cutting the trees which give us oxygen to breathe, let us stop using fossil fuels which are causing climate change. Let's start learning, and re-learning how to be in sync with nature, and not just using nature, for a sustainable future.

Let's start this movement, let each one of us take the responsibility of correcting the mistakes that we ALL have contributed to, knowingly or unknowingly.

Let's do it, NOW:

Let's play our role by taking every single tiny and big step to make the world sustainable.

Let's take a pledge to avoid the use of energy, as much as possible.

Let's be a disciplined user of energy.

Let's take a pledge to minimize the use of energy as much as possible.

Let's use only efficient appliances.

Let's take a pledge to generate and fulfil our energy needs using locally available resources.

Let's establish Energy Swaraj at every level.

Let's make AMG (Avoid, Minimize and Generate) our guiding principle towards the adoption of Energy Swaraj.

LET'S ESTABLISH ENERGY SWARAJ – LET'S START ENERGY SWARAJ MOVEMENT

Modelled on the lines of Gandhiji's Gram Swaraj which propagated self-reliance in production and consumption of goods for the day-to-day needs, Energy Swaraj propagates self-sufficiency in energy. Energy Swaraj is the localised energy generation and consumption to bring in energy self-sufficiency at the household level, community level, in villages, in cities, in metros, in countries...

...all over, across the world.

The generation and consumption of energy locally, using local renewable energy sources, which with current technological advancements is very much feasible and economically viable, provided one is sensitive to the energy needs. If we become a disciplined user of energy, generating and using only when required, then with the current technology, we would be able to generate and fulfil all our every energy need on a 24x7 basis.

Therefore, through this book, I would like to make an Appeal to ALL for adopting Energy Swaraj.

My appeal is not for a few people, but for every household, every institution, community, societies, offices, factories, industries and governments all across the world.

I believe an initiative like this requires nothing short of a movement to make it effective, as the climate change is happening very fast, hence the corrective measures should also match the pace.

All individuals need to come forward to join this Energy Swaraj movement, adopt 100% to Solar energy and declare *Energy Independence.* You need to take responsibility and become a crusader of climate change and messiah of green and clean fuel for the Earth's sustainability.

As argued earlier, we should not expect the Government to take such a step where we think of complete localised energy solutions. We, ourselves need to think of complete independence from energy and sustainability of life on the planet in the long-term. Hence it is required that individuals take this initiative and join the Energy Swaraj movement.

The technical and economic viability of Energy Swaraj has been discussed in earlier chapters.

MY APPEAL

TO THE YOUNG GENERATION: *MAKE SOLAR DEVICES YOUR MOBILE*

Though you are not responsible for it, unfortunately, you are going to pay for it. Climate change is going to affect the young generation the most, who are going to live longer on this planet than anyone else. Therefore, my first appeal is to all youngsters. Firstly, as you play with your mobile phones, start learning to play with Solar or renewable energy technologies. You can start making your own Solar devices and more importantly start using them. If you are a student, start learning to make your own Solar lamp and start using it.

If you are reading this book, you are most likely be living in a house which is fully electrified. Take the first step, make your own Solar lamp, switch-off lights and start studying under a Solar lamp.

Once you have reduced your energy consumption, then start asking your parents to minimize their energy consumption and eventually generate all of it using Solar energy.

TO THE HOUSEHOLDS – *BRIGHTEN, LIGHTEN YOUR HOME WITH SOLAR ENERGY*

I believe the most important stake-holders of the Energy Swaraj movement are households, all across the world. I strongly appeal and request your contribution to this movement.

In order to adopt complete Energy Swaraj, I appeal that you surrender your electricity connection, minimize your energy needs and then start generating entire energy using Solar energy. Surrendering electricity connection may be difficult for people living in multi-storied buildings, but I am sure that everyone with their own rooftop can do this.

If you can't do this immediately, you can start taking small steps to save energy. To achieve this, you can consume as little energy as possible. Become sensitive to the use of energy or its wastage. Slowly, step by step, you can switch to 100 per cent localised energy solution.

If people across the world start practising this, then it will become a great movement. And, without Energy Swaraj becoming a movement, no major impact will be created on the mitigation of climate change.

TO THE FARMERS: *PLAY A KEY ROLE IN SUSTAINABLE LIVING*

Sustainability is all about producing and consuming in a sustainable manner. The farmers feed everyone in the world. If farmers start producing food by sustainable means, using less water, fewer pesticides, using Solar energy, it will make the world a better place.

I appeal to all the farmers of the world, big and small, to become sensitive towards the means of sustainable living and sensitise the world. Adopt Energy Swaraj, generate and fulfil all your energy needs in your farms, in your house.

TO INSTITUTIONS: *BECOME SOLAR ENERGY WARRIORS*

I appeal to every school, college, university, company, NGOs to join the Energy Swaraj movement. You can ensure that your institution is taking all the measures to first, Avoid the use of energy, then Minimize use of energy, and then Generate it using renewable energy.

What could be a better place than to have educational institutions to initiate the movement? These are the temples of knowledge and with this, they can add another feather in their cap and become warriors of Solar energy.

What could be a better place than the workplaces, where we spend a significant and active portion of our lives and help to create the world in the way it is today. Why can't each workplace join the Energy Swaraj movement and create a world that is sustainable? Why can't each of these workplaces, across the world, become sustainable themselves?

TO GOVERNMENTS: *IMPLEMENT STRONG POLICY MEASURES*

It would be very difficult to think of converting, nearly complete centralized energy generation to completely decentralized generation, which is required to establish Energy Swaraj. However, strong policy measures should be implemented to start the transition from nearly centralized generation towards nearly decentralized energy generation.

Even, when there is a need for centralized generation, probably to power metro cities and big industries, the government can adopt large scale Solar and Wind power plant with energy storage facilities.

In general, people are living everywhere on the planet, renewable energy resources are also available everywhere on the planet, therefore the sensible approach would be to generate energy wherever people are living. Please do not treat Solar and Wind energy solutions as asset creation and asset management. Yes, it is easy for the government to achieve scale in this manner and fulfil the targets, but it misses out on the bigger picture.

The benefits of adopting Energy Swaraj will provide tremendous benefits to countries. Therefore, I appeal to governments across the world to adopt Energy Swaraj for job creation, for local skill development, for creating assets in local economy, for empowering women and children, for reducing dependence of people on external energy, for reducing dependence of countries for external energy and for improving energy security of the country. This approach can provide you with innumerable benefits.

JOIN THE ENERGY SWARAJ MOVEMENT – SURRENDER YOUR ELECTRICITY CONNECTION

Finally, my appeal, once again, to ONE and ALL - Energy Swaraj Movement can only be a global movement if driven by people. I cannot imagine any other shape of the movement.

Let's all work together towards minimising energy consumption, give up usage of fossil fuels and to achieve where all of us generate and fulfil our energy needs.

Let's take pride in Energy Swaraj, let's write and sing songs and poems on Energy Swaraj.

Let's put a flag in front of our house, institution showing that we are energy independent.

Let's feel proud that we are not polluting or damaging the environment, thus mitigating climate change, by adopting Energy Swaraj.

This is not an ordinary situation; this is an emergency. In an emergency, the 'business as usual' approach does not work and the success of economies is not the only criteria. Overcoming current dangers of climate change and establishing sustainability would require extraordinary efforts.

Let's do whatever it takes.

Let's put extraordinary efforts.

Let's adopt Energy Swaraj, for the self, for the community, for the region and for the country.

Let's be part of this movement – Let's establish **Energy Swaraj.**

I am appealing to each and every one of you to reduce your consumption of energy and then switch to 100% Solar energy.

I am appealing 100,000 families to surrender their electricity connection by 2nd October 2020 and make a valuable contribution towards saving the climate and the planet.

A not-for-profit, non-governmental organisation called Energy Swaraj Foundation has been established with the motto, "Solar is Life". Energy Swaraj Foundation would be ready to help you in every manner for switching to Solar.

The organization will help you to provide a solution or will help you to connect to the nearest solution provider.

Visit www.energyswaraj.org and take a pledge for surrendering connection.

Be the first. First 100,000 families or institutions who surrender their electricity connection will get a special mention in the Energy Swaraj Movement.

The Energy Swaraj Movement is ON!

Signing off for now!

SAVE OUR EARTH

Have you ever taken the time
To see what was around you?
Taken the opportunity
To gaze at every little detail of your surroundings?
Or have you closed your eyes stubbornly
Because you prefer to be blind?
In the truth
You don't want to admit that our precious earth
Is falling apart
It's beautiful skirts of grass
Is dying
It's garden of stunning plants
Is running out of air
And us,
Us, the ones who made it this way
Will one day perish into a dark hole
In a way,
We are all selfish murderers
Killing the sol we walk upon
With our trash and disgrace, our ungratefulness
Slowly
Slowly it shall fade
Slowly we shall perish
Save the earth

Save our hearts and souls
Or forever be lost
In a pit of death and regret
Save our home
Save our air
Save our lives
Save Our Earth

– Jessica Robert

REFERENCES

1. www.wikipeadia.com

2. www.cia.gov

3. Carbon Dioxide Information Analysis Center (CDIAC)

4. World Resource Institute

5. Carbon Brief

6. Earth System Dynamics

7. www.mkgandhi.org

8. CDIAC, Our World in Data

9. Earth System Dynamics., Vol-7, pages:327–351, Year-2016):

10. UNDP

11. HDI data

12. www.nationalaffairs.com

13. www.oneworldeducation.org

14. www.ourworld.unu.edu

15. IPCC – Sustainable Development Goal

16. www.nationalgeographic.com

17. www.seametrics.com

18. www.conservation.org

19. www.earth-policy.org

20. www.globalagriculture.org

21. Gram Swaraj by Mahatma Gandhi

GLOSSARY

1. **AMG:** Avoid Minimize Generate
2. **Anarkali:** Courtesan in Great Mughal Emperor Akbar's Court
3. **Bharatnatyam:** A dance form originated in Tamil Nadu
4. **Biomass:** Organic Matter used as fuel
5. **CFL based Solar lamp:** Compact Fluorescent Lamp
6. **Chai Thela:** Tea Cart
7. **Chajjas:** Cover of roof
8. **Charkha:** A Spinning wheel for weaving khadi
9. **CNG:** Compressed Natural Gas
10. **CSR:** Corporate Social Responsibility
11. **Dharma:** Righteousness
12. **Ekal Vidyalayas:** Single teacher schools
13. **Gram Swaraj:** Self-reliance in villages
14. **Green-house effect:** Natural process that warms the Earth's surface
15. **Karma Yogi:** Saint who believes in noble deeds
16. **Kulfi:** Frozen dairy dessert made of milk
17. **LPG:** Liquid Petroleum Gas
18. **MP:** A state in Central India
19. **NCPRE:** National Centre for Photo-Voltaic Research and Education
20. **OMG:** Oh My God
21. **Pagdandi:** Narrow passage for walking in the fields

22. **Pilot Project:** Small – Scale preliminary Study
23. **PV Module:** Photo-Voltaic Module
24. **Prarthana:** Prayer
25. **Satyagrah:** Protest for the truth
26. **Swaraj:** Independence / Self-reliance
27. **Tabla:** A membranophone percussion instrument
28. **Verandah:** Courtyard
29. **Zille – Subhani:** Salutation used for Great Mughal Emperor Akbar the great

CPSIA information can be obtained
at www.ICGtesting.com
Printed in the USA
BVHW031908041019
560273BV00001B/56/P